D1808647

The JOHN A. LIVINGSTON READER

THE

John A.
Livingston

READER

The Fallacy of Wildlife Conservation
AND
One Cosmic Instant

McCLELLAND & STEWART

Copyright © 2007 by the Estate of John A. Livingston
Appreciation © 2007 by Mysterious Starling Inc.

The Fallacy of Wildlife Conservation first published in 1981 by
McClelland & Stewart Inc. Copyright © 1981 by John A. Livingston

One Cosmic Instant first published in 1973 by McClelland & Stewart Limited.
Copyright © 1973 by John A. Livingston

"The Prophetic View of John Livingston," from *Rescue the Earth!: Conversations
with the Green Crusaders* by Farley Mowat. Copyright © 1990 by Farley Mowat.
Reprinted by permission of the author.

All rights reserved. The use of any part of this publication reproduced,
transmitted in any form or by any means, electronic, mechanical, photocopying,
recording, or otherwise, or stored in a retrieval system, without the prior written
consent of the publisher — or, in case of photocopying or other reprographic
copying, a licence from the Canadian Copyright Licensing Agency — is an
infringement of the copyright law.

Library and Archives Canada Cataloguing in Publication

Livingston, John A., 1923-2006
     The John A. Livingston reader : The fallacy of wildlife conservation
and One cosmic instant: a natural history of human arrogance / John Livingston.

Includes an essay by Graeme Gibson.
The fallacy of wildlife conservation and One cosmic instant previously
     published separately.
ISBN 978-0-7710-5326-6

     1. Environmental ethics. 2. Wildlife conservation. I. Gibson, Graeme,
1934- II. Livingston, John A., 1923- . Fallacy of wildlife. III. Livingston,
John A., 1923- . One cosmic instant. IV. Title.

GF75.L58 2007   304.2'8   C2006-904257-8

We acknowledge the financial support of the Government of Canada through the
Book Publishing Industry Development Program and that of the Government of
Ontario through the Ontario Media Development Corporation's Ontario Book
Initiative. We further acknowledge the support of the Canada Council for the Arts
and the Ontario Arts Council for our publishing program.

Typeset in Filosofia by M&S, Toronto
Printed and bound in Canada

In *One Cosmic Moment*, portions of Chapter Two, "One in a Million," appeared
in different form in *Canada* by John A. Livingston (Natural Science of Canada
Limited, 1970).

This book is printed on acid-free paper that is 100% recycled,
ancient-forest friendly (100% post-consumer recycled).

McClelland & Stewart Ltd.
75 Sherbourne Street
Toronto, Ontario
M5A 2P9
www.mcclelland.com

1 2 3 4 5   11 10 09 08 07

CONTENTS

APPRECIATION

by GRAEME GIBSON

Almost forty-five years ago the families Livingston and Gibson moved into adjoining sides of a semi-detached house in Toronto's Rosedale neighbourhood. John was the executive producer of science programs at the CBC. I was teaching English at the Ryerson Polytechnical Institute and was almost three years into my first novel. We both had young families, although mine was younger than his. It didn't take long for me to be won over by his combination of charm, intellect, honesty, and bear-like energy. I was soon calling him Jake, and from that time on he most often spoke to me as m'boy. At some point along the way, I began to feel that he was as much an older brother as a friend.

What I remember best from that time are the dinners, and what dinners they were! With between eight and ten friends crowded around our dining room table, the Livingstons always amongst them, these meals featured great standing prime rib roasts of beef (at sixty-some cents a pound), with glasses of whisky, and, most often, Hungarian Bull's Blood red wine. By the end of an evening we'd all had too much of everything and were either arguing or singing, and sometimes we managed both. It may have been on one of these occasions that Jake asked me if Beethoven's Ninth Symphony justified the human extirpation of a single species of bird, the peregrine falcon, for example. I cannot imagine that either of us expected me to answer that, and I certainly didn't try. It was merely a step in an unspoken exchange that was happening between us. In retrospect I suppose he'd already begun to teach me.

The early sixties were a busy, melodramatic, and ominous period. Newspapers maintained their circulation nicely with the Bay of Pigs invasion and the Berlin Wall, with the Cuban Missile Crisis and the Space Race, which was parading national egos and gobbling money. Nuclear bomb tests blossomed on every side, and even Vietnam raised its head with Kennedy sending in his first group of "advisers." It isn't surprising that most people I knew at the time had had at least one nuclear war dream.

Obviously, there were other kinds of news stories as well, but most of us paid little attention to them. Some dealt with an alarming series of wildlife die-offs resulting from the commonplace spraying of chemicals such as DDT. In 1962, Rachel Carson's iconic book, *Silent Spring*, finally propelled these concerns dramatically to the surface. Symptomatically, I was scarcely aware of its publication. I was deep in a hard and seemingly endless struggle with my first novel, dreaming of another kind of apocalypse and picketing the American and Soviet consuls after each new testing of the Bomb; my imagination was seized by another book published that year, Günter Grass's *The Tin Drum*.

John Livingston was the first bird enthusiast I'd met, and I found the activity puzzling: the common view of bird-watchers back then had them as little old ladies of both sexes, in sensible raingear and with funny hats – rather like train spotters, but without the trains. However, Jake didn't fit this mould; he was an assured and forceful man, one who loved food and drink, big band swing, traditional jazz, and the Hamilton Tiger-Cats. And yet he was fascinated by the manner in which our own domestic cats hunted co-operatively: his nonchalantly diverting the feeding pigeons from a distance while mine crept in for the kill. And how, when his old tom once brought in a vole, rare in the Toronto area, his surprised and enthusiastic response to the gift had caused the animal thenceforth to drag home every conceivable variety of small mammal it could find. But above all, for him, it was the birds: he obviously loved them best.

It took me some time to succumb to the birds, but I finally did about three years later, when a very large hawk unexpectedly soared out from beneath the Bloor Street viaduct. On impulse I went to a bookstore and purchased Roger Tory Peterson's *A Field Guide to the Birds of Eastern and Central North America*. When that wasn't enough to identify which hawk it was, I bought binoculars. It turned out to be a red-tailed hawk. By then I was hooked, and began to seize every moment I could to explore the Rosedale Ravine or wander along the lakeshore, where I first discovered there are more than two species of duck.

It amazed and humbled me to find that I'd lived more than thirty years without ever seeing the extraordinary vitality and beauty of the bird life around me. How could I have missed the astonishingly delicate multi-coloured warblers, the fly-catchers and dashing hawks, the woodpeckers and various thrushes that I was painstakingly learning to identify in the centre of Toronto? The flute-like spiralling song of a wood thrush in the ravine where I'd walked for years now stopped me dead in my tracks. Surely it had always sung there at this time of the year? Why hadn't I heard it?

Birds had been around for over 150 million years by the time early humans arrived on the scene. We came to self-awareness as a species surrounded by them. Almost from the beginning they have played critical roles in myth, religion, and therefore art. Birds are one of the great and ancient lineages of life, and yet, like most of us, I scarcely knew they were there.

It gradually became clear that, as a result of human-centred preoccupations, I was out of touch with the non-human world. What I encountered every day, what I dwelt in every day, was an elaborate cultural construct that had very little to do with nature. That is to say, I discovered a truth that is axiomatic to all John Livingston's writings. It was undoubtedly also at the heart of the life he lived.

Early in *One Cosmic Instant*, Livingston reminds us that the World's Fair at Montreal, Expo '67, was subtitled *Man and His World*, and that its amusement-park successor remained *Terres des Hommes*. The implication of this naming demonstrates a principle underlying Livingston's thought, and, for that matter, the reality of our human situation. In short, he maintains that our dysfunctional and profoundly destructive relationship with the natural world is an inescapable result of our belief that the human animal is separate and different from everything else in nature. The corollary of this is our overweening conviction that everything on and in the earth is ours to do with as we please.

At the time of his death, John Livingston, who had earned his B.A. in English in 1943, while on active service in the Canadian Navy, was Professor Emeritus in the Environmental Studies graduate program at York University. He was a natural teacher, admired and highly influential. Ex–graduate students describe him as inspiring and supportive, bold and refreshing, a man whose moral compass was set just right. There is palpable admiration and affection in the way he is remembered. *Yfile*, the York University bulletin, said after his death that the Faculty of Environmental Science had been "flooded with messages from people wishing to honour and respect Livingston's life and achievements. The many stories circulating about him clearly demonstrate that his influence is deeply felt and his legacy will continue to inspire people all over the world."

He was president of the Canadian Audubon Society (now Nature Canada) from 1955 to 1962, during which time he edited *Canadian Audubon* magazine and directed the society's research and information programs. He was particularly concerned with the protection of "problem" species such as the whooping crane, the sandhill crane, and various hawks and owls, and with oil pollution at sea.

He also served as president of the Federation of Ontario Naturalists. He was a founding trustee of the Nature Conservancy of Canada, and a founding partner in LGL Ltd., Environmental Research Associates, where he studied arctic birds for the Berger Pipeline Inquiry. A former member of the education commission of the International Union for the Conservation of Nature, he also served on the board of the Rachel Carson Trust, and those of various other nature preservation organizations. From 1962 to 1968 he was the first executive producer of the award-winning CBC Television series *The Nature of Things*, to which he contributed until the

1990s. Between 1966 and 1994 he published ten books, about which I will speak shortly.

John Livingston was also a member of an extraordinary generation of Canadian thinkers. The years following World War II were a period of remarkable critical energy in Canada: in the twenty-two years between 1947 and 1969, Northrop Frye's *Fearful Symmetry* and *Anatomy of Criticism* were published, as were Marshall McLuhan's *The Mechanical Bride*, *The Gutenberg Galaxy*, and *Understanding Media*, and George Grant's *Philosophy in the Mass Age*, *Lament for a Nation*, and *Technology and Empire*. These three writers were all innovative structural thinkers whose work in one way or another addresses what we are in danger of losing, or indeed have already lost. I suspect that the experience of World War II's great technological and human violence influenced them all.

Livingston's *One Cosmic Instant: A Natural History of Human Arrogance*, *The Fallacy of Wildlife Conservation*, and his final book, the Governor General's Award–winning *Rogue Primate: An Exploration of Human Domestication*, fall within this structural tradition. Like the works of Frye, McLuhan, and Grant, these books deal with a fundamental re-examination of the implications of our human relationships with nature. And Livingston was certainly preoccupied with what we have lost.

As a thinker, Livingston is closest to George Grant, whose abiding concern was the conflict between traditional Western values and the new religion of technology. However, where Grant argued with melancholy passion that our historical notions of justice and personal freedom are grievously threatened by technological progress, Livingston insists that these historical values are at the heart of our destructive relationship with nature. Nevertheless, the two men had much in common:

in *One Cosmic Instant*, Livingston emphasizes Grant's observation that "When one contemplates the conquest of nature by technology one must remember that that conquest had to include our own bodies. Calvinism provided the determined and organized men and women who could rule the mastered world. The punishment they inflicted on nonhuman nature they had first inflicted on themselves."*

I suspect, too, that Grant would have sympathized with Livingston's sense that quality of life "lies in the richness of the life experience *within* flows and forms and processes that are not of human manufacture . . . [and] as the richness and variety of the environmental whole are reduced, simplified, and degraded by human activity, so the richness and variety of the individual human life experience are proportionately affected."

It is interesting that three of the four thinkers I have mentioned graduated in English Literature – Grant's discipline was religion – and that Rachel Carson started out in English before switching to zoology. At least in Livingston's case, one is tempted to think of John Ruskin's assertion that birds should not be studied by scientists, but rather by artists. Or as Thoreau wrote in his journal, on March 23, 1853: "Man can not afford to be a naturalist, to look at Nature directly, but only with the side of his eye. He must look through and beyond her. To look at her is fatal as to look at the head of Medusa. It can turn the man of science to stone."

When asked by a young man working as a master bird-bander whether he should go back to university and study biology, Livingston advised him, if was going to do anything, to complete his degree in English instead.

---

* See page 350.

The titles of Livingston's books alert us to what's in store. *One Cosmic Instant* and *Arctic Oil* are respectively subtitled *A Natural History of Human Arrogance* and *The Destruction of the North?* His final book, *Rogue Primate*, is subtitled *An Exploration of Human Domestication*. *The Fallacy of Wildlife Conservation* scarcely needs elaboration.

One Cosmic Instant is perhaps the best analysis of how and why it is that humans have historically disconnected themselves from nature. The argument is intelligent, passionate, melancholy, and richly informed, as was the man himself. It is also tough. Set against this argument is a movingly eloquent celebration of the richness and astonishing complexity of the life from which we have relentlessly distanced ourselves. Crediting Leo Marx in *The Machine in the Garden*, Livingston says that man has set himself outside nature, and only observes it and derives pleasure from it. This illusory separation of mankind from nature is in large part the result of what Livingston calls "the dream of immortality," which he believes to be at the heart of our human alienation. Livingston would refine and develop this conviction for the rest of his life. He explains it in the following passage from *One Cosmic Instant*:

> The flutes of Arcady were never there, but then, neither was there original sin. There was no "noble savage," there was no Yahoo. There was only introspecting, self-deceiving man, frantically searching for homocentric meanings, purposes, and designs, when all the time there were none. The enormity of that simple truth has yet to dawn on us.[*]

The Fallacy of Wildlife Conservation, as a statement of belief, is one of the fiercest and most uncompromising of John

---

[*] See page 345.

Livingston's convictions. Had he entitled it *The* Failure *of Wildlife Conservation*, we might have tried again — without having to think too much about it. But he didn't. Titles, or subtitles, were always key to his work. As a result of the word *fallacy*, we are confronted with an insistence that we rethink everything. The implications of this are both intimidating and, for many, demoralizing; moreover, it is far from clear whether we are capable of conceiving of a more compliant and therefore less self-centred way of living on the earth. As he says himself in *Arctic Oil*:

> Alternative ways of doing things are frightening enough, but alternative *perceptions* of things are downright terrifying. Perceiving goes on inside our individual heads, and we have a very deep, vested interest in what we receive into inventory. . . . The missionary is just as afraid to let dancing pagan goblins enter his personal universe as his victim is to part with them.[*]

And yet, in *The Fallacy*, he's telling us that we will not stop or even significantly control our assaults on the non-human world until we change. We have to change, because, as he says in *One Cosmic Instant*, "There is no engineering answer to a problem created by culture." I'm not sure that John himself ever managed to see how this change could be accomplished, which perhaps explains the tragic quality in his voice. He tried to see, of course, and in that striving we find some of his most affecting writing.

Responding to Livingston's position by saying he's too pessimistic, as some have done, is like saying that a man's arguments are compromised because he's too short. Optimism

---

[*] John A. Livingston, *Arctic Oil: The Destruction of the North?* (Toronto: Canadian Broadcasting Corp., 1981), page 114.

and pessimism are merely attitudes towards possible outcomes; neither makes an argument more or less valid. Undoubtedly, there were both optimists and pessimists in the years leading up to the fall of Rome or the collapse of the Hapsburg Empire. I once heard the fine English painter Francis Bacon asked in a BBC interview why he was so pessimistic. "I'm not," he said. "I'm a flaming optimist." Then after a pause he added: "There's just nothing to be optimistic about." John Livingston might well have agreed.

At the heart of John Livingston's value system is his conviction that man is not innately savage, because nature is not; that morality is in our genes; and that, tragically seduced by our dreams of immortality, we have taken the wrong path as a species. Accompanying this is his rejection of the notion that evolution has anything whatsoever to do with competition, or "survival of the fittest," as market fundamentalists or Social Darwinists would have it.

In place of savagery and competition he sees The Will to Comply. "I'm not trying to be 'mystical,'" he writes in *The Fallacy of Wildlife Conservation*, "in suggesting that after a lifetime spent in the appreciation of nature, the single factor that appears to me to be more important than any other is *compliance*. I can very comfortably interpret ecologic interdependence as co-operation. When a predator takes one of 'my' grouse in the bush, it is not a crime against me or against God. It is co-operation on the part of the grouse."*

Most social animals share the will to comply, and a most compelling demonstration of it can be found amongst wolves. Should an upstart younger animal challenge the alpha

---

* See pages 89-90.

male and discover in the process that he isn't up to it, he throws himself puppy-like upon the mercy of the leader by lying on his back, or else offering the most vulnerable part of his throat. Compliance here goes both ways, as it must: the young animal submits utterly, and the alpha wolf doesn't attack. Thus an ancient social order is maintained, which makes absolute sense if one considers the grievous damage done to a group of wolves if one or both of their most capable animals is injured in a savage fight.

Konrad Lorenz took a similar lesson about compliance from the wolves, one that enlightened him by clarifying a passage from the Gospel that had always irritated him – "And unto him that smiteth thee on the *one* cheek offer also the other. . . " (Luke 6:28). We aren't being asked to do this, suggests Lorenz, so that our enemy can strike us again, but to make him *unable* to do so. This is precisely what Livingston means by saying human decency does not originate in our culture, but in our genes. In the Introduction to his *History of Warfare*, John Keegan makes a similar point: that men have to be coerced into performing the kinds of violent destructive acts that constitute warfare. They don't do these things "naturally."

Neither a politician nor a missionary, John Livingston does not pretend to have clear answers to our present human predicament. Indeed, one of the more unsettling elements in his work is the possibility that, given the reality of life on earth, there may not be any such answers. He suggests that one of our more peculiar characteristics is the conviction – indeed the insistence – that there be solutions. Because we are convinced there is a reason for everything, humans have come to believe that all problems can be logically addressed and resolved. However, Livingston admits that he has no solution to the problem of wildlife preservation. "There may not be one," he says, "but given the somewhat shaky assumption that one exists, I sense that I can at least feel the direction." He then suggests that the direction

> seems to lie in the compliant acceptance by individual human beings of membership — which is to say, 'place' — in the beauty that is life process. This is too easy for me to say, for the difficulties are formidable. Such an act — a conscious and willing act of self-recovery — would require the recognition of options that have long been masked from us by our accumulated tradition. It would require the healing of perceived dichotomies, the mending of conceived alienations. . . . It would be the dissolution of the ancient western divorce from nature.*

Although this is pretty dire stuff, his logic rings true. And yet, Livingston behaved as if there were indeed productive practical things to be done, apart from this total change in human consciousness.

And so there are, at least in the short term. I suspect that on some level, John Livingston also believed this. As a teacher, writer, filmmaker, and conservation activist, he challenged several generations to acknowledge the extreme gravity of what we are doing to life. By insisting that the root of our current environmental crisis is our delusion that we humans are something separate from — and *better* than — animals, Livingston lays the onus on each of us as an individual to make an inner change. Although technology and politics ameliorate the problem, they will not solve it. It is up to us.

> As every wildlife preservationist knows to his cost, the simple-minded among us are wont to interpret our anger as misanthropy. I try to tell them that we are honest folk; if we were

---

* See page 151.

misanthropists, we would say so. Still, it is often charged that we 'care more for animals than for people' – again an unwitting emphasis of the 'man/nature' problem. This, as I perceive it, is cultural constipation at its fullest. It is impossible for such as these to shift – even momentarily – outside the cultural hallucination to see that what we care about is life – all life in all forms – or, perhaps better expressed, living. I care for the phenomenon of living more than I care for the contorted belief systems of my culture. I care for wildlife preservation more than I care for its opposite – the technomachine that serves it. I care most deeply of all about the failure of wildlife preservation in my lifetime.[*]

Inherent in his work and his life is John Livingston's act of faith that, despite our relentless depredations, humans are incapable of entirely destroying life on earth – that its extraordinary complexity and energy will survive us as a species. There is much solace in this, especially for a man whose arguments – clear, intelligently presented, and tough as they may be – are all informed by the reverence, the almost spiritual love he felt for the beauty of life and the sublime authority of all of its processes.

That some form of mysticism, or intuitive experience – call it what you will – is close to the centre of the ethos Livingston shares with most naturalists and preservationists is indisputable. At the same time, as Livingston emphasizes, the spirituality of traditional religions or of individual "mystical" experience has failed to develop as a force, or even as an effective presence in the struggle on behalf of non-human nature. There's a great pity in this, for should we manage to tap the demonstrable power of collective spirituality, it may well provide our best hope.

---

[*] See pages 149-50.

When someone asked George Grant about his pessimism, he apparently replied: "I'm not being pessimistic at all. I think God will eventually destroy this technological civilization. I'm very optimistic about that." Jake might have agreed with this sentiment as well.

The PROPHETIC VIEW *of*

JOHN LIVINGSTON

An INTERVIEW *by* FARLEY MOWAT, 1990[*]

John Livingston is a contemporary of mine, and we have had much in common as we struggled to understand the role mankind plays in the theatre of life on earth. We have pursued the same objectives, but followed different paths. Whereas I became something of a preacher, John developed into a phenomenally adept teacher and the prime philosopher of the environmental movement in this country.

The clarity of his insights, the breadth of his grasp of what ails this planet, and the ruthlessness of his conclusions combine to give him the stature of an Old-Testament prophet. Like one of those, he is revered by true believers but feared and rejected by those who are preoccupied with worldly self-interest.

---

[*] From *Rescue the Earth: Conversations with the Green Crusaders* by Farley Mowat.

F.M. John, what brought you to an awareness of your indivisibility with other living beings?

J.L. There's a phenomenon which takes place in pre-adolescence that if one is exposed to other species – it doesn't matter particularly whether it's the dog or the cat, but better still if it is multi-species – it results in a connection which is never lost.

Paul Shepard says youngsters literally ingest the experience of nature and it becomes part of them. I believe that this is so. In my case, it happened to be a fascination with newts and toads and frogs and stuff when I was a small child. Then I got interested in birds. I was a keen birder all my childhood.

F.M. What was your first intimation that all was not well between man and the natural world?

J.L. When I was a youngster in Toronto they decided to put a storm sewer through *my* ravine – Cedarvale – which meant ripping the heart out of the place. I remember weeping with rage, anger and frustration because it felt as if part of me was being cut out. It was like a piece coming out of my stomach, and I was only ten or twelve. I may have been a bit "bent" by that experience. But from then on I was committed to defending nature.

My first involvement in conservation work came in the early fifties when I was named executive director of the Canadian Audubon Society. That thrilled me because of all those magic names of great naturalists I had never met. Roger Tory Peterson, Alexander Sprunt, Alan Cruickshank were all on the Audubon staff at the time. It was as though heaven was here.

F.M. How lucky you were to have found a place so early on.

J.L. It *was* a place for me, and I mean "place" in the most profound sense, both social, physical and ideological.

I was involved in a lot of things there but most especially the protection of birds of prey, because they weren't protected under the Migratory Birds Act and were shot on sight as vermin by sportsmen and farmers alike. I spent an awful lot of time lobbying individual provinces to try to get them to enact protection for hawks and owls. Unfortunately the demands made on me by the Audubon head office in the United States were rather different from saving wildlife – they expected me to concentrate on administrative and financial affairs. There was a heavy debt to be realized, and I wasn't the guy to do that. So in 1962 I went to the CBC and began *The Nature of Things*. I was the first producer, and I stayed there until 1968. Then I went to York University where I've been teaching ecology ever since.

F.M. When did your involvement with Pollution Probe come in?

J.L. The founders were my friends but I was never formally aligned with it. By that time I already suspected that the seat of the mischief was somewhat deeper than pollution. And, too, I was developing a terrible problem dealing with institutionalized beliefs, which had been so manifest in the Audubon Society way of perceiving and doing things. The conservation movement seemed to be becoming an alphabet soup of conflicting ideologies, none of which, I felt in my heart, had much to do with my frogs and toads and newts and warblers. By the mid-sixties I was becoming convinced that the roots of the difficulties between man and nature were cultural, as opposed to blaming government, industry and other such specifics. That's when I went to York University.

A university is a great refuge for a person like me. One can also make a modest contribution. If one can turn students on by slicing the onion in a different way, then one has done something useful. All of my work since 1969-70 has been in attempting to come to grips with and resolve the conflict between man and nature as I happen to perceive it.

F.M.   You don't like being called a conservationist any more. Tell me why.

J.L.   I'm *not* a conservationist. Shall we debate the semantics?

F.M.   Aren't conservationists those who try to protect and preserve life on earth?

J.L.   That isn't conservation as it is currently understood and applied. In practice, conservation is now resource development. The World Conservation Strategy holds that conservation is resource-oriented, and can only be defined in terms of human utility – in effect saying that everything was created by an almighty beneficence for human use. Therefore conservation becomes management – the "wise use" of these things in the long-term human interest.

F.M.   It's generally short-term.

J.L.   The rhetoric is long-term. The *fact* is short-term – economic horizons being what they are, and political horizons being even shorter.

F.M.   In other words, the original meaning of conservation has been perverted to mean exploitation of nature for human gain?

J.L.   Of course. Maximum economic return is what the word has come to mean. Conservation is resource development under intensive management. It is the antithesis of *preservation*.

The "preservation" meaning of the word is anathema from a technocratic point of view. As far as they are concerned, conservationists are those who use nature to pave the road for the technomachine, whereas preservationists are exemplified as stupid old ladies with their hearts on their sleeves who squat in the path of orderly progress. Preservationists are not to be taken seriously because they don't share in the ethos of the necessary primacy of human enterprise.

F.M.   What do you think this book I'm writing is about, if not about conservation?

J.L.   It looks as if it's at least partly a review of the institutionalized approaches to the perceived loss of natural resources, more than the loss of nature. But even the loss of nature is proprietary. *Our* birds, *our* endangered species, *our* woodlands, *our* swamplands are disappearing. *Our* natural heritage – I despise that attitude. Nobody in the respectable conservation movement is talking about nature for her own sake. Only individuals like you and me are weeping solitarily for nature for *her* sake.

It's hard to say whether the terrible sadness I feel is for the ongoing destruction of nature or for my own loss of the experience of nature.

F.M.   I don't think you have to differentiate; it's all one.

J.L.   I don't think so either. If anybody hurts my dog he hurts me. So both are damaged. That's not an easy concept to institutionalize in the technocratic society.

F.M.   Isn't that what the animal welfare movement is trying to do?

J.L.   Yes. Some of them. People like Stephen Best, Vicki Miller and Anne Doncaster are trying to articulate this, and are not embarrassed to say this. But no hard-line, middle-of-the-road conservationists would ever dream of uttering such things because of their terrible anxiety to be thought respectable and credible.

F.M.   Read: rational and logical. Are they afraid of emotional, subjective attitudes because such things are not fashionable in a mechanistic world where the intellect has become the godhead?

J.L.   There is a terrible paranoia in the neo-conservation community. Even within naturalist organizations, maybe especially these, there is paranoia about being seen as being emotional, and this fear has allowed them to mesmerize themselves right into the technocratic orbit. In order to appear respectable they permit themselves to become pawns of the Ministry of Natural Resources or whatever.

F.M.   Are they being co-opted or are they co-opting themselves?

J.L.   Both. Why don't naturalist groups, especially, have the guts to stand up and take a position against sport hunting and fashion fur? That would be too emotional. Can't do that, because we've got to play ball with the bureaucrats. This is the part that hurts more than anything else, that organizations begun by people with a deep and abiding love of nature are becoming indistinguishable from the conservation bureaucracy itself.

   The only movement left that's speaking from the heart is the animal welfare movement.

F.M.   What about Professor Peter Singer, the guru of animal rights? He is a very logical and rational human being – or appears to be.

J.L.   He would almost professionally deny being an emotionalist because he has to be a proper philosopher. Don't forget that logic is technology. How to think, how to address a problem, how to define a problem, how to argue. The whole structure is so tyrannical, and it's part of technology. The philosophers get very angry when I say that they are part of the technomachine.

F.M.   What *about* the business of hunting? Why *shouldn't* people hunt for sport?

J.L.   I can't get around the idea of people getting their jollies by maiming and wounding and killing. Hunters deny they do this, of course. They say it's just like the bird watcher getting out and enjoying some kind of communion with nature. People have written whole books about hunters communing with nature. They say a mystical experience is derived from this bloodletting; which I simply cannot understand. "Man *is* a natural predator" is the rationale that's usually produced in scholarly circles in defence of sport hunting. They say we evolved as predators. I deny this. We evolved as gatherers, with a little bit of hunting on the side. The hunting instinct is no more ingrained in us than any other aberrant behaviour. The simple fact is that hunting has become an industrial enterprise – a very big one and a very big source of profit to any number of people. There's also the macho aspect in that most hunting, now, is carried on by urban folk because it's a "U" thing. Fortunately, hunters are becoming a shrinking minority, albeit a very vocal and powerful one. I think sport hunting is on the way out in spite of the hype it gets.

I once had to write an article about hunting, and I went to the local newspaper shop and bought the current issue of all the hunting magazines and read them all. What I found most informative was the advertisements. The language and style of the advertisements was so overtly sexual it was hard to believe.

A study Statistics Canada did a few years ago has the most compelling evidence I've seen that sport hunting is becoming passé. It was a review of the perceptions and uses made of nature by the general public – how many people like to watch birds, how many like to garden, how many like to go out in nature, hiking, visiting a conservation area, all that. Less than ten per cent said they liked to hunt! I suspect sport hunters are a dead-end expression of human domestication, of bad breeding, like the pit bull terrier.

I think it always was abnormal behaviour. It was the fullest expression of the commodification of nature. Animals are there to kill, and mount on a wall, or photograph yourself beside, and this is the ultimate form of consumption.

F.M. What about subsistence hunting? Don't you see any justification for that?

J.L. We've got a lot of historical revisionism going on these days about that. Subsistence hunting – what I call subsistence in its pure form – is ecologically circumscribed. You are hunting for your life. No problem there. But now the fur industry and its apologists in government and in several conservation organizations have redefined subsistence to include commercial hunting and trapping.

It's absolutely indefensible. There's no ecological, anthropological or economic basis for saying that the fur trade is part of the traditional aboriginal culture. None. It was the fur trade and the trade in animals of all kinds that *destroyed* the aboriginal cultures of North America. And to hear people today, including natives, claim the anti-trappers are going to bring down the native culture is bizarre.

I think the great change to the aboriginal cultures came when the native no longer saw the animal as part of himself, part of his environment, but as a source of cash. The animal was transformed into a cash symbol. The moment that happened, the aboriginal culture was gone.

One knows why they are making these statements now – because it's a part of the argument for land claims, and if one can claim one needs the land for aboriginal cultural purposes, then it's another little bit of support for one's case. However, proprietorship of land is just as foreign to aboriginal cultures as the taking of wildlife for cash.

F.M. What happens if we convert the wild-fur business to so-called fur ranching?

J.L. Captive breeding. I believe that the suffering of the captive-bred animals is infinitely greater than that of the wild-caught animal, because a wild animal shows very little psychological response to being attacked and killed if the action is done swiftly. But captive animals suffer over the whole duration of their lives. They live out their lives in terrible apprehension. These are not domesticated animals; they are captive *wild* animals, especially mink, weasels and foxes. If there is anything wilder than those three predators, I'd like to know what it is. They all go crazy when you confine them. They can't be domesticated. The essence of domestication is social dependence, and there's no social dependence for them.

F.M. Would you say that a lot of the effort being expended against trapping and hunting might be better used elsewhere? That these problems are going to look after themselves?

J.L. Oh no. Commercial hunting and trapping will go on long after amateur recreational hunting has gone. So the money and effort is being well spent.

F.M. But aren't you being pretty hard on native trappers?

J.L. There's no money in it for the natives anyway. The money is all made at the wholesale, retail end. Native trappers average $200-300 in a year. It's aberrant behaviour for people to be wearing the furs of living beings for ostentation, for status symbols, as prestige goods.

F.M. Like having lampshades made out of tattooed human skin?

J.L. I don't think it's essentially any different. Can you imagine this sort of thing happening in nature? It is anomalous behaviour right from the top down. I see it as anti-social in a fundamental way because I firmly believe in inter-species social organization. I am absolutely convinced that inter-species relationships are the ultimate relationships. We start with individual selfishness, then our relationship sphere begins to expand. There's mother; there's family;

there's the tribal self, and that eventually transcends to an inter-species self. At that stage, the essence of the feeling one has for nature is selflessness. The individual self dissolves in the overall relationship. A participatory relationship with nature. I don't want to stand outside and be a Cartesian observer; I want to participate, to feel. A man by the name of Morris Berman wrote a wonderful book called *The Re-enchantment of the World* in which he rejects the machine-like world of scientific objectivity. His goal is the retrieval of nature as sacred, as enchanted and as participatory at all levels, individual, group and so on. This is the kind of thing I am interested in.

F.M. How do we get rid of the mechanized world we have constructed and recapture awareness of where it is that we belong?

J.L. I know it sounds like a cop-out, but all I can say to that question is I have to understand it better and more deeply before offering a solution.

F.M. It may not require any conscious solution. The mechanical world may simply destroy itself, if not with nuclear devices, then through a collapse of the life-sustaining envelope as a result of our tinkering with it.

J.L. Leo Marx wrote a wonderful book called *The Machine in the Garden* about an image that originated long ago in Europe of the garden and the machine that kept invading and trying to subjugate it. True conservationists have always been out there fighting against the machine. This is his thesis. I don't pretend to comprehend why we should feel that way, except I suspect it is the pure expression of instinctive self-preservation.

I speculate sometimes with colleagues that true conservationists are analogous with the spruce budworm; it is the spruce budworm's job to prevent the overriding dominance by the spruce trees over the rest of the living community of the forest. Dominance in too great a degree makes for greater vulnerability of both the dominated *and* the dominator.

F.M. Technology *is* dominance for the human species, and it is clear that every technical advance since the invention of the hand axe, or before, has made us increasingly dominant, but also increasingly vulnerable to massive catastrophe by making us dependent on our artificial constructs. Our vulnerability as increasingly alien beings on Planet Earth has grown almost geometrically as technology has enabled us to become ever more dominant.

J.L. And so we have become increasingly dependent for survival, not on nature but on an external agent which we have ourselves created. Technology.

F.M. Perhaps we are becoming as vulnerable in our reliance on technology as is the cosmonaut in his tiny mechanical planet.

J.L. The space analogy is a good one. Technology is about nothing if it isn't about expertise, and narrow, compartmentalized expertise at that. Define the problem in terms of a solution is the name of the game. I think your average technocrat is a bladder stuffed with solutions, looking for ways to turn these into power. If the problem doesn't have a solution, then redefine the problem to fit the solution at hand. That is the essence of technocracy and of the cultivation of expertise. Matters of the heart and the being and the bone marrow don't have much to do with expertise. And expertise without these qualities of *living* things isn't going to solve the problem of how to save this planet.

Everything hinges on *preserving* — not conserving, mark you — quality habitat, whether it holds aquatic or terrestrial communities. Everything derives from quality habitat, and that means plants — without which there can *be* no animals.

F.M.   Habitat protection seems now to be the main thrust of many conservation organizations.

J.L.   It is – and that's good. But in many cases it still doesn't mean preservation – it means management. And what does that mean? It means sustainable development.

F.M.   Setting aside "natural" regions to be used for a different kind of exploitation?

J.L.   That's right. I don't like the expression "set aside." It's like dessert that we'll set aside for later consumption. The rhetoric of many neo-conservationists appals me.

F.M.   There is a whole new language here. A kind of newspeak.

J.L.   And one has to understand it. "Sustainable development," "maximum sustained yield," "resource conservation," "wise use" are in fact attempts to sugar-coat our ongoing intention to continue to exploit nature for our own, often indefensible, ends. Human utility remains the name of the game. I think many so-called environmental organizations have deliberately set out to demonstrate the utility of nature to the human enterprise as the most palatable way of selling conservation. And it has become so ingrained with them now to argue the cause of nature in terms of human requirements – real or otherwise – that they can no longer escape from it. To see conservation organizations like the World Wildlife Fund, the Federation of Ontario Naturalists, or others who fundamentally stand for the interests of nature – to see them argue in terms of human utility nauseates me, because it is caving in to the demands of the technostructure. Instead of the consumer or commodifier having to *prove* the legitimacy of claims on nature, those who would defend nature have to prove a case for its protection. Unfortunately, the argument they use is inescapably utilitarian. This is a suicidal intellectual as well as emotional position. And this is what I dislike the most about the institutionalized conservation movement.

F.M. If you stand only on the defence in war, you court inevitable defeat. What we should be doing is demanding of the users or abusers of nature: "By what right do you do this?"

J.L. This is correct. And in this connection I have to add that "animal rights" – as opposed to "animal welfare" – is a stupid argument in my view because there is no way to argue rights outside the human enterprise. So by arguing for animal rights we are again anthropomorphizing nature and going on the defensive.

We are imprisoned by rational arguments. This is the part that hurts. You hear the most impassioned, eloquent arguments of all reduced down to nature's utility to man. Even the aesthetic is a utility. Even the appeal to preserve nature for future generations is a utility. Leave it for future generations to utilize – that's what it boils down to. Somebody has to have the guts to fight for nature simply because it's in our marrow, and because we love it and can't live without it!

It is so sad to see conservation groups falling into that trap, becoming instruments of the very organizations and governments they are trying to oppose.

F.M. I wonder if there is any real alternative. Taking the subjective and emotional point of view automatically disqualifies you from a respectable hearing. Nobody is going to listen. So maybe you have to fight them with their own self-serving logic.

J.L. If you do, you're dead. And in the end, the living world will be dead too.

The

FALLACY

of

WILDLIFE

CONSERVATION

CONTENTS

NOTE *to the* READER

Although the preservation of wildlife is both my bias and my vehicle, this essay is really about the difference between living and being, which is sometimes called "quality of life." It was written in response to a direct challenge. For years I had been uncritically mouthing the conservation catechism; it was time to think it through. What follows is an account of that process.

ONE

The BACKGROUND

I am not a biologist, an ecologist, nor indeed an "ologist" of any kind. I am merely a naturalist, for whom logic and "the word" have come to count for little. Nature, praise be, neither talks nor is rational, and therein is comfort.

Naturalists become accustomed to travelling light but, still, there are the accumulated impedimenta of a lifetime; in a little pouch of talismans I carry such things as the scent of fresh tiger dung in the monsoon jungle and the stench of two million puffins on fogged St. Kilda. There is the loving rasp of tundra saxifrage across my palm and the shocking sear of man-of-war tentacles around my chest. I once ate a malarial mosquito distended with my blood and I can summon up the after taste of fried musk-ox cheek. My bones have felt a hyena I could not see, crushing a wildebeest skull, and I have celebrated, with a cock peafowl, crowing in the night. Odours, touches, tastes, sounds – all part of what at this moment is me. This is not to say that I have not *seen* much – I have – but one learns that visual traces can game at chance with the memory;

not only are pictures illusory, but also seeing is believing, and believing can be a one-way street.

I am a conservationist, a nature preservationist, and I have smelled the exhausts and the people and their works, heard the slogans and the guns and the exhortations, felt the nylon and the concrete, tasted the saccharin and the tenderizer. I no longer dare to *look*. The last time I looked, there was an endless tumbling cataract of identical managers, administrators, and technicians. There wasn't a freedman in sight. There were a few splashes of pigeon whitewash on a massive bunker that houses an institution of higher training and, within it, coveys of tight-lipped souls with pocket calculators and hard briefcases bent on "managing our world's resources for the future" through rational planning, design, and development. Next time, *you* look, I'm busy ruminating.

A ruminant is a cud-chewer — a cow or a goat. To extract optimum nourishment, the ruminant chews and swallows its fodder, then brings it up later for further chewing at leisure. The human ruminant, as exemplified by the videotape highlight analyst or the academic, is a re-hasher. As it is for the hoofed being, re-hashing for the human being can be a highly beneficial exercise. In light of our peculiar human specialization, perhaps for "re-hash" we should read "re-view." In this essay I want to review a few things about wildlife and conservation and some related matters.

One way or another, I have been involved in wildlife conservation all of my adult and a good portion of my sub-adult life. It follows, if only by definition, that I have known frustration deeply and long. There are all manner of frustrations in conservation, of which one of the more obvious would be that which arises after you have laboured long and assiduously in the vineyard only to have somebody fly along and dump herbicide all over it. But most of our problems are subtler.

No abrasion stings more than that which is caused by one's own carelessness, stupidity, or indeed innocence. No bruise is more tender than that which is one's own fault. Indeed, perhaps no resentment is deeper than that of which the source is directly traceable to oneself. We in the wildlife conservation cabal have enjoyed a few successes, but failure and futility have become our familiars, and we are hurting. Our overwhelming losses can instantly, easily, be attributed to the urban-industrial process of scarification, to world human populations, to institutional greeds, and all the rest of it. But such excuses are much too facile. To lay the total blame here, I feel, is a species of Afghanistanism: railing and shaking your fist at an unseen foe who is quite unaware of your existence, much less your fury. Better to stay closer to home. The fault is not and never was "out there."

Not that we have not worked and thought hard; we have. It is true that we have always lacked the financial muscle to play the political game where that game is played, and that we have lacked the ideological empathy of the majority of the "general public," but these are no more than the facts of our existence. Disadvantage is simply the environment of conservation; it always will be. But there are many admirable beings who prosper in apparently wholly inhospitable surroundings. See a baby Lapland longspur crack its natal shell warm and pulsing in a June arctic blizzard; see a yellow wood sorrel crack asphalt; see a gosling or a turtle hatchling leg it across the sand to the sea amidst a flurry of gulls or frigate-birds. Our "hostile" environment is slim excuse.

I think that a depressingly large portion of the wildlife conservation effort has been entirely wasted. In view of the fact that in the conservation energy budget there is no margin for waste, no subcutaneous fat for emergencies, this is all the more alarming. If I am even partially right, then it might be well to review our ineffectiveness, to isolate even a few of the more obvious contributing factors. I think this is a worthwhile exercise. We have nothing to lose.

Defensive action – delaying action – is always terribly busy and reflexive and reactive, simply because there usually is not time in which to regroup, dig in, consider, and strate-gise. Confusion and fragmentation – and exhausting flailing – often follow. Such would be my characterization of wildlife conservation: we dart about, stamping at tiny smoulders in the carpet, rushing from hot spot to hot spot, when all the while the roof is racing to a fire-storm and the walls are creaking toward collapse.

People in the "line" of conservation fire-fighting (there is nothing you could call "staff") have rarely had time to draw back and take a painstaking look at what we are actually doing. Each of us no doubt waxes choleric over both the priorities and the pathetically limited ranges of action of the organiza-tions we support, and each of us has a personal short list of critical hot spots. Which will it be – polar bear, Kirtland's warbler, clouded leopard, fin whale, tuatara, Sumatran rhino? But I am not going to get into that. I want to speak in more general terms about some questions that very rarely make it even to the absolute bottom of the agenda. Some practising conservationists may not want to hear about them at all; for a long time I did not. Now I find them inescapable.

I suspect that our difficulties may be more entrenched and obstinate than many of the more "activist" among us appear to realize. It is one thing to raise placards in the street and another to put the aggregate messages to content analysis. That is what I will try to do here. No doubt many will see this as the most reprehensible cop-out – avoidance of action. At a younger age, I was repeatedly instructed by John H. Baker, the late distinguished president of the National Audubon Society, to keep my shirt on, for the love of God. I choose to believe now that if we sit down, catch our breath, wipe our spectacles, and walk – not run – back to the mêlée, we may yet extend the day.

In the course of this pause for reflection, I want to address not the practice but the *argument* for wildlife conservation. Of course there is no single argument – they are many and varied – but all involve reason and logic of one sort or another. I want to look at that reasoning process. In conservation we have always assumed a dialogue between ourselves and everyone else; a civilized, adversary proceeding in which reason, logic, and meticulous argument, liberally laced with horrible precedent, would persuade just men and women to our position. We have invested enormously in that assumption. Unfortunately for reason and logic, for ourselves, and for wildlife, it has not worked. One would like to know why.

What follows cannot be exhaustive. I do not pretend to know all of the factors that are involved. I shall stick to those with which I have had some experience. Suspecting as I do, however, that we have been effectively arguing *against* ourselves, in the first part of this essay I shall look at some of the "stock" or "set-piece" arguments, and try to weigh their utility and identify some of their implications. Then I shall try to abstract some of the more deeply embedded malignancies that, perversely enough, seem to thrive and prosper on our arguments much as forest, field, and garden "pests" respond to biocides. Finally I wish to speculate – admittedly in a very personal way – on possible alternatives or at least on supplemental ways and means.

This search is important for several reasons. After spending more years than I really care to count in trying to advocate the *practice* of wildlife conservation – the "doing" – I became interested in trying to understand the fundamental *notion* of wildlife conservation – the "knowing." Wildlife conservation not necessarily as an activity, but as a state of mind. Because it is always so easy to set down in little square boxes someone *else's* motives, and to impute and attribute and assign perceptions and beliefs to others, much thought was then invested in describing the peculiar state of mind of the opposition (still assuming an adversary proceeding).

I have written and lectured on the historical and cultural roots of a state of mind pervasive in the western world (but not unique to it) that appears to be implacably indifferent to wild nature. The western cosmology would appear to be so firmly man-centred that the very notion of wildlife conservation can barely be entertained in any profound sense. Much of the conservationist's frustration arises when, even though his argument appears to be apprehended intellectually, it is obvious at the same time that the emotional or spiritual shift from the cool abstract to the hot-flowing, pumping urgency of *living* is for some reason impossible. It is as though the necessary "leap" into our own biological context cannot even be conceived, much less entertained.

But even abstract intellectual grasp of the argument is rarely achieved. There are few experiences quite so cruel as, when after having made a well-argued, even elegant and moving case to someone, you lean back expecting the warm sparkle of shared insight and understanding, only to encounter the flat opaqueness of complete and utter incomprehension. The lecturer whose delightedly prepared joke flops wetly flat knows how the wildlife conservationist feels. As anyone who professes for a living has to learn sooner or later, if one's little academic jests develop a consistent pattern of failure, then perhaps it might not be wise to invest too much energy in sifting for faults in one's audience, but rather in going over one's own material. That is why I have set about trying to understand our own array of arguments for conservation and to assess, as best I can, their effectiveness and worth, their actual result. That is not easy. Anyone undertaking such a task, whether stand-up comedian, evangelist, or soothsayer, knows that cleaning up one's act can be an untidy and often painful process. I have as deep an emotional and practical investment in this question as any reader, and what follows may not always be flattering. But again, we have very little to lose. Just as I write this, a new modulation in the song of a magpie-robin outside tells me we may have something to gain.

Like it or not, we shall first have to deal with definitions. Such a procedure can often be dull, but in this case it is vital. One reason is that the words "conservation," "ecology," and "environment" have been so widely and so cynically co-opted in our time that they barely retain a shred of their original meanings. Another reason is that I am weary of playing semantical games of the jesuitical sort so adored by the humanists, who are no longer biological beings, but of some other cosmic stuff. This is simply to say that the ground rules here are going to be those of wildlife preservation.

Very generally, by convention, "conservation" has meant the care of "natural resources" and their protection from depletion, waste, and damage, so that they will be readily at hand through perpetuity. Traditionally we have used this approach for soil, water, vegetation, and, in a more particular way, to those kinds of animals (fishes, fur-bearing mammals, waterfowl, and the like) in which there is a commercial or other special interest. More recently we have begun to see "sustainability" in some endangered plant and animal species and associations. For these, the long-term goal is not necessarily exploitation, but rather preservation in the cause of what are considered to be somewhat higher human aspirations and values than sheer consumption. In fact, the more recent and interesting of the wildlife conservation arguments tend to coalesce around these latter aspirations and values, but in order to understand their genesis and evolution, and their implications, we must review those justifications for conservation that many of us might consider more crass.

(Quite recently, you will have noticed an increasing tendency to separate out and indeed to polarize "conservation" and "preservation." Many hunters, developers, planners, managers, and others will proclaim themselves as being conservation-oriented in the most modern, aware, and realistic way, by contrast with the old-fashioned, stubborn "preservationists" who squat sullenly and stupidly in the way of orderly progress. This peculiar conceptual dichotomy is obviously as useful to the production-consumption parade as it is destructive of wild nature. Preservation smells of reaction, retrogression, primitivism, and worse, while environmental assessment, regional planning, ecological development, and all the other appurtenances of the technomachine are permitted, through "conservation," to assume their proper heroic roles.)

Traditionally, "resources" (including wildlife) have been perceived in an entirely utilitarian light. Someone has observed insightfully that resources do not exist in and of themselves; they *become*. The moment we see usefulness in something — anything — that thing becomes a resource. Resources are human assets. Natural resources are those assets that were put in place by natural processes, such as trees or uranium, and that we can draw upon as required, for such purposes as the extractor (and by extension, society) sees fit.

We have drawn a distinction between "renewable" resources and those, particularly minerals, that are "nonrenewable"; once you take them out of the ground you cannot replace them, or grow more of them. At one time, this seemed a reasonable approach, but in our own day it may become difficult to maintain. More on that later.

By definition, a "resource" of any kind has *intrinsic utility*. If we could not use it, then it would not be a resource. This applies equally to "renewables" and "nonrenewables," including wildlife. The notion of *use* is central and nuclear. At one point we began to make a further distinction, in this case between "consumptive" and "nonconsumptive" uses of renewable resources, including wildlife. Consumptive uses of wildlife would be fishing, sealing, whaling, sport hunting, and the like, in which the individual being is "consumed" (killed). We cannot use it without killing it. Whether we literally consume (eat) the animal is irrelevant here; we have taken it out of circulation for good. In contrast to this, we have such "nonconsumptive" uses of wildlife as birdwatching, nature photography, or skindiving, in which, although we *use* animals, we do not kill them. This once-useful distinction has also become blurred.

But we see in wildlife and other "resources" more than utility. Just as important, perhaps more so, is the unstated but implicit assumption that the use, whatever it is, is vested solely and exclusively in the human interest. The ownership and the proprietorship are ours alone. The balance sheet is the human balance sheet, and assets whether current or fixed are man's inalienable property. Thus man has private ownership of and vested interest in diamonds and anchovies, topsoil and snow geese, duckweed, balsawood, and sperm whales. Human beings have absolute, divinely conferred right of access to these "resources" for whatever human uses may be contrived. The fundamental concept of "resource" is clear and unequivocal.

So it is that when we draft a conservation rationale, almost without exception, we argue the case in terms of the human interest. It does not matter which of our several constituencies we may be addressing. At the simplest level there are identifiable special interests in agriculture, forestry, fisheries, mining, recreational fishing, and shooting. These are obvious. Also we have the special interest of science and technology: clearly we have an investment in the preservation of the raw material of knowledge, with all of its implications for research, education, technical application, and so on – the advancement of human affairs in general. Finally we have the apparently growing body of people who have an interest in the preservation of nature for aesthetic, ethical, or even spiritual reasons. But even this interest is no less "vested," in the sense of implicit payoff. There is a future benefit to us.

The definition of "wildlife conservation" that I use daily, and from which I am working here, is this:

> The preservation of wildlife forms and groups of forms in perpetuity, for their own sakes, irrespective of any connotation of present or future human use.

By "wildlife forms and groups of forms" I mean living beings and groups of living beings (associations) existing in a natural state: plants, animals, and their habitats. I explicitly exclude from this definition: man and other domestic and feral animals and plants; species introduced by man and living in foreign environments; and man-dependent species. This last means that as long as there are people around, there is no need to expend any effort preserving rats and silverfish. In essence, wildlife conservation is the preservation of nonhuman beings in their natural settings, unaffected by human influence or activity, uncontaminated by human antibiosis, emancipated from human serfdom.

I hasten to acknowledge that there are few places on planet Earth that are not at least indirectly touched by man and his accomplishments. I am very much aware that besides being idealistic, my definition has the usual taint of proprietorship — or at the very least, paternalism — that seems to characterize most of our attitudes (intellectual-emotional positions) with respect to nonhuman nature. I trust you will forgive me, on the grounds that each of us, more is the pity, is no more than human and we must communicate in human terms; that there would be no need for "conservation" — itself a wholly human concept and activity — were there no human creature; and that we ourselves created the vacuum into which the need for conservation flows, and nobody else is going to do anything about it.

Whatever your individual perception or definition of what wildlife conservation is, or might be, our collective track record is spelled out in nauseating detail in many a readily available place; I am not going to recount it here. Most historical accounts of conservation activities tend to arise from the special perspective of whoever happened to write those accounts, and thus are just as biased as my perceptions and yours. There are many perspectives; take your choice. If you happen to be interested in the captive rearing of birds, you will have been much rewarded by the resurrection of the Hawaiian goose, and you will be watching the current breeding programs for whooping cranes and peregrines. If you happen to be a duck-shooter you will be able to point to the many habitat-restoration programs your organizations have undertaken. (If you were to take a slightly closer look, you might see that you have not left enough ducks to fill the habitat available to them.)

If your interest is endangered species in general you will be watching efforts toward specific reserves and legislative controls of exploitation and destruction. (If *you* look harder, you will see that the 'ologists and legislators cannot yet decide what an endangered species *is*!) If you are a park or zoo enthusiast you will have noticed the recent proliferation of both. If you are an ethologist you will be pleased by recent spectacular advances in your science, this time among wild animals. If you are a television viewer you will have come to know an endless assortment of "nature-adventure" programs. On the other hand, if you are that unidentifiable "impartial man-in-the-street," you may very well have become a little jaded by the omnipresent multichannel trumpetings of "environmentalism," which whatever it is should not be confused with wildlife preservation.

Whoever you are, whatever your special or even peripheral interest, it is very difficult to back away from immediate specifics for a wider-angle view. Each of us who has the remotest concern for wildlife is at least a kind of specialist. Each of us has places, landscapes, species, habitats, problems, that especially preoccupy us. Each of us is cranky about *something* a little more than the rest. I happen to like the desert, the tropical forest, and the arctic more than most other places; I loathe cities and love the sea; I prefer predators to herbivores; I like birds of any size more than small mammals; I prefer non-game to game species; I am pleased about symbiosis and bored by taxonomy. I like mysteries. There is no profound significance in any of this; it is merely the way I am. The way each of us is, the way each of us sees, and feels, and responds.

So when I invite you to join me in zooming back a little for a broader perspective on wildlife conservation I do so in full awareness of the bias that shapes and directs my own reflections. Nevertheless, I think that I should make the effort. And if I can, you can. When you do, I am certain that you will come to the same conclusion: in the broadest sense, wildlife preservation is a catastrophic, heart-breaking disaster.

The dismal figures are widely published and well-known. The terrifying trends are readily visible. I will not detail any of it here save to remind you that the torpedoed Ark is settling with all hands — varieties, races, populations, species, associations, communities, whole faunas — and almost without a sound.

Remember that as the absolute number of organisms, species, and groups of species diminishes, the likelihood of recovery is geometrically rather than arithmetically reduced. Also, since the crucible out of which these vanishing miracles were formed is *all of Earth time*, and the devastation has taken place in no more than a wink of *human* time, things are in no position to take care of themselves. The wonderful mutualistic, compliant, selfless, all-encompassing orchestration of relationships which is in process of brutal eradication can never be reformed, rebuilt, reassembled, "restructured." By comparison with the biosphere, Humpty-Dumpty is child's play.

Entirely out of control, the human technomachine guzzles and lurches and vomits and rips its random crazy course over the face of the once-blue planet, as though some filthy barbaric fist were drunkenly swiping with a gigantic paint roller across an ancient tapestry. Unevenness in the rich textured nap of Earth's surface causes the paint to cling slightly unevenly, with scattered spots and holes showing in the roller's wake. These are the success stories. Isolated and discrete as they are, it is quite possible that they can never recombine into a coherent whole.

We are left with a miscellaneous rag-tag assortment of odd and disconnected relics — some larger, some smaller. In general, these gaps or anomalies tend to be in "frontier" regions (arctic, rain-forest), or in the great biological near-desert that is the open ocean. But the mindless machine has long since outgrown all restraint; the paint roller's antibiotic lacquer is thick, and fluid. If you watch, you can see its viscous pools widen, as though of their own volition, toward the farthest reaches of life's lovely tapestry.

Wildlife communities are richer, in terms of numbers of species, in equatorial regions than in higher latitudes. The greatest number of endangered species in the world today live in the tropics and sub-tropics. In underprivileged over-populated countries I see no glimmer of hope for wildlife. It is simply too much to expect, for an entire catalogue of reasons, that the care and maintenance of wildlife could possibly rise on the list of priorities (including accelerated industrial expansion) that exists in the tropics today. We should remember that there is little or no preservation tradition in most such places, in any event, and to think that such a tradition could spring forth fully developed in the face of current events would be to abdicate common sense altogether. The human orgy has exactly one conceivable outcome for those species that are (a) edible, or (b) compete with man for food, or (c) compete with man for space. Perhaps this will change one day, but not soon. In the meantime, losses will have been colossal and extinctions will have been many. Extinctions without replacement – ever.

On the other hand, there are places in the world that are not yet populated toward the point of ignition. There are some spots not yet painted over, not yet obliterated. In spite of their technical grotesqueness, some of the hypermanaged western nations still have options having to do with open space, natural areas, living nonhuman beings. But even there, we have to admit the utter failure of wildlife conservation either as a practice or as an ethos to penetrate the general consciousness. The acceleration of wildlife extermination is remorseless, even in the "civilized" world – perhaps especially there. There is a general almost total failure to grasp the *notion* of wildlife conservation.

Here I mean societies such as the one I live in, where one might expect the cultural environment for the notion to be wholly favourable. Places where the "haves" live. I do not, by the way, mean regions or nations influenced or dominated by French, Italian, Portuguese, or Spanish traditions, which have never been favourable for wildlife or its protection. On the evidence, wildlife seems to hold its own best in "have" cultures deriving from northern Europe, including Britain, poorest in those deriving from southern Europe. The implications of this must surely be obvious to all but the most doggedly unobservant.

The worst prospect for wildlife is in those countries where human population increase is out of control and/or in those that have inherited the "romance" traditions. And political ideologies have nothing whatever to do with it. When you think about this, it dawns that ownership of the means of production and locus of the distribution of profits are entirely irrelevant to wildlife conservation, as long as the goal of human activity is production. Who cares who owns the whaling fleet, the automobile factory, the petrochemical plant, the jet aircraft assembly line? Who cares where the profits go? What earthly difference does it make to wildlife? Certainly it has never made any difference to conservation.

On a world basis, "wildlife conservation" in its fullest and deepest meaning as "preservation" simply does not exist. That is because its fullest and deepest meaning cannot be expressed in a political platform, a computer printout, an official plan, or a research report. You cannot quantify, analyze, show data, and prove it out. It is not like that. For me, wildlife preservation is a wholly permeating life awareness that has become an unconscious part of every thought, attitude, perception. So it is too with others, many of them; but not *that* many. Not enough to matter.

Then one must ask, why *aren't* there more of us? People in the mass can be and are every day persuaded to the most bizarre assortment of freakish, perverse, weird, and convoluted intellectual positions the craziest of minds can concoct. What is the matter with *us*? Certain (as are all the converted) that there is nothing unhinged about our position – that life is manifestly *right* and that the wilful reduction of life is *wrong*, why are we, the unique possessors of THE WORD, so pathetically inept in adding to our number?

No doubt I will be challenged on this point. Much of the conservation rhetoric calls attention to the growing number of people who have become involved, in one way or another, with our "movement." I have done so earlier in this chapter. That the number is growing in an absolute sense I have no doubt, but then so is everything else growing in an absolute sense – all except wildlife. In a relative sense the story is dismal.

When, for example, you state a case for wildlife to any tier of government, you are told there simply is not sufficient mandate for action of the magnitude that is clearly needed. Insufficient mandate means inadequate constituency. One learns to judge politicians on grounds other than their utterances, but in this case they are absolutely right. We wildlifers like to *claim* that there is a constituency – or several of them – for wildlife preservation, but the harsh and unequivocal record shows that we are neither statistically nor politically significant.

There are abundant reasons for our political failure. First, of course, is the fact that in spite of our widely advertised "environmental" awareness, the polis is the human concentration and admits no other participatory interest. Also, the political tenure system is in polar conflict with ecologic realities; life systems take their own deliberate time, for which new promises in new elections cannot wait. Political process and nature preservation are fundamentally antithetical. And, as I have pointed out already, political ideologies are simply not relevant.

But there is more to our cultures than politics, and there are reasons for our failures that are much less obvious. Indeed there are so many of them, so interwoven, intractable, and complex, that one despairs of ever beginning to do an adequate job of sorting them out. That is why, in this instance, I have chosen not to look so much at our "opposition" but rather more at ourselves.

I do not suggest that our arguments are *the* cause of our failure — they obviously are not — but I shall try to show that they may have played a much more important role than many of us may have realized. In any case, to examine one's own logical process can only be salutary.

Methodologists urge us to begin with a parsimony of assumptions. Mine are rather too many to allow a hyperclinical approach, but no matter. By acknowledging those of which I am aware I may at least begin to expose some of the dimensions of the problem. For purposes of what follows, then:

I assume that wildlife preservation is a good thing.

I assume that there is a rational argument for wildlife preservation.

I assume that whatever that argument is, we have not yet used it effectively.

I assume that argument *can* be used effectively, which is the same thing as saying that I assume that reasonable people respond to reasonable argument.

I hypothesize that those arguments we have used have been unequal to the task.

TWO

The ARGUMENTS

SELF-INTEREST

If we can't be good, at least we can be prudent. You do not strangle your own support system. If for no other reason than our own welfare, our own selfishness, we should try to entertain at least the possibility of some slight merit in a conservation-oriented approach to living. This is the core of the self-interest group of arguments. You might go so far as to say that if we cannot appreciate the nature and quality of our acts and anticipate at least some of their consequences, then we have no right to describe ourselves as intelligent, or even sane.

In some contexts this approach works. Not many years ago it would have been unthinkable – preposterous – to suggest that people might actually band together to prevent flights of supersonic aircraft over large cities; or stall the building of airports; or haul mercury, lead, and asbestos poisoners into court; or redirect highways, dams, pipelines, and other such narcissistic exercises. But these have been done, and in the name of self-interest.

These of course are not wildlife issues; they are people issues. There is a world of difference. Let us look at some of the ways in which self-interest is expressed in terms of wildlife.

These are the oldest of the lot, and they appear in a number of varieties. It will help if we isolate a few of the more prevalent of these before considering "wise use" as a whole.

Husbandry

If you treat whatever you are using intelligently, you can help to assure its continuing usefulness. This idea has long been basic in the handling of livestock, ground covers, topsoils, and so on. Good husbandmen have known it forever. They have known that their interests lay in the interest of whatever they were cultivating, whether cattle and crops, or the soils, water, and vegetation that sustain them. You find the best of husbandry, in the pure sense, in the breeding and rearing of domestic animals, where it is both a science and an art. It seems to involve not only meticulous care but also a kind of special insight that is perhaps no more than an indefinable "feeling" in the experienced stockman. A sort of touch that some have, but most of us have not.

The notion of husbandry was extended to natural as opposed to domesticated resources many years ago, and for very good reason: rural people could understand and appreciate its implications. Whether applied to stands of trees, water sources, watershed slopes, or fertile valley bottoms, good husbandry meant care and sensitivity in their treatment. Such was in both the individual and collective interest.

Husbandry of course can properly apply only to something that is under your total control — your herd, your flock, your vineyard. If it is not under your total control you cannot husband it. Nevertheless the notion was eventually applied to certain forms of wildlife – also with good reason. Those wild animals chosen for "husbandry" were those desired either for commerce or for "sport" (which in the present day is a very real form of commerce). By treating the species concerned intelligently, as a good husbandman, you ensure its continuance in your special interest.

Such, essentially, was the rationale for the first closed game seasons, bag limits, and so on. Later, the regulatory style became much more refined, as in minimum sizes for fishes or "males only" for deer. Still later, wise husbandry dictated the repair, maintenance, and even the construction of suitable breeding habitats for most-desired ducks. There appeared fish hatcheries, transplantations of foreign game birds, wild animal farms, all aspects of what later came to be the massive manipulation of the wildlife "resource." An enormous complex of detail emerged: kinds and sizes of nets for commercial fishes, different bag and possession limits for different species of ducks, and so on. All of it "wildlife husbandry," in the interest of the resource user or consumer.

The notion of "wildlife husbandry" is inherently and ridiculously self-contradictory. Animal husbandry's most important element is selective breeding. If wildlife is wild, then it cannot be husbanded. But it sounds good, no doubt. In an apparent attempt to keep the illusion of husbandry alive, those charged with the delivery of wildlife to the consumer (in North America, chiefly provincial and state departments of fur, fish, and game) chose to describe their work as wildlife "management." This was singularly appropriate, as it tended to anticipate the propriety – indeed the sanctity – of the religion of management in the technocratic age.

There are serious implications. Practitioners of wildlife "management" are almost entirely concerned with the uninterrupted production and mass packaging of commercially important species. Indeed, many contemporary wildlife biologists are frustrated that their charge is so rarely extended to those species (meaning the vast majority of species) not commercially desirable. Although there are some efforts, chiefly at the federal levels in Canada and the United States, work on behalf of non-commercial wildlife is relatively minute in both budget and priority allocations.

More significant are the implications of the husbandry argument as addressed to the public. Emphasizing as it does a few desired species to the almost total exclusion of the overwhelmingly greater list of native plants and animals, it suggests that the former *are indeed* the more important species. If it is to these that virtually all of the effort is directed, then clearly that must be an accurate reflection of the proper priorities. How else is an individual ignorant of wildlife to interpret it? Every day of our lives you and I must accept someone else's judgement on everything from television programming to foreign policy because there are supposedly qualified people making decisions for us on the basis of information we do not have. So it is with "wildlife husbandry." The experts know what they are doing; who am I to question it?

But the problem is deeper. Even assuming that the public were to become more sophisticated about wildlife conservation needs, there remains the matter of the assumed fitness and rightness of good husbandry, intelligent treatment, enlightened management. The important assumption, as the argument is drummed into us by both government agencies and private organizations, is that all of wild nature is a herd, a flock, a crop, to be manipulated and controlled in the public, national, and human interest.

Man is total proprietor, manager, and decision-maker with respect to wildlife. This is the clear-cut and unambiguous message of "good husbandry" conservation. Wildlife is yours; yours to manipulate in your own best interest. If you treat it badly or stupidly, only you (not wildlife) will be the loser; if you treat it well, it is yours from which to profit in perpetuity.

Stewardship

There is more involved, as it turns out, than individual and collective management for profit or loss. The true husband-man, almost by definition, is sensitive to the continuity of things. His knowledge, his insight, and his skills are not vested in him once only and to the exclusion of others. He recognizes both an indebtedness to the past and an obligation to the future, if only in respect to the welfare of his science. He sees himself as temporary proprietor of the herd only, having inherited the wisdom and accumulated experience of generations gone, and carrying the responsibility of refining that experience, to the best of his ability, toward its further refinement when his time is over.

This is the good steward personified. He sees himself as no more than short-term custodian, performing an honest and honourable task that in its total scope transcends the individual human lifetime. Thus it is that the appeal for conservation is directed to an entire generation of such custodians. Or, on another plane, it is addressed to all of mankind in respect to the biosphere itself, in the service of a higher spiritual power.

The ultimate stewardship entreaty, which casts man in the role of resident manager on behalf of the Creator, is for many people a reasonably palatable interpretation of Old Testament injunctions having to do with man's "dominion" over non-human nature. Taken literally – as they frequently are – the

Biblical pronouncements crash and reverberate to the drums and jackboots of absolute tyranny. Interpreted, however, as direction to a kind of benign caretaking, in the name of God, with the resident manager responsible and answerable to God, the instruction becomes tolerable to many people.

Indeed, if for "God" we were to read "life process," or "biosphere," or "cosmic continuity," or some such, there would seem to be relatively few of us who could not act comfortably within this stricture. That it does effectively confer absolute power and authority — if only temporarily — on the custodian is not seen by most as too grave a problem because with that power and authority go responsibility and answerability. Clearly nothing could be in the better interest of the temporary custodian than to do his job wisely and well. In this way the good husbandman is serving an even higher interest than the human collective in the here and now, and even in the future.

This refinement of the "husbandry" argument is similar with respect to the seat of power and authority in the relationship between man and the rest of nature. Even though we are in the divine service, on Earth we are unquestionably in charge. If we treat resources badly or stupidly, we are not only the losers in the sense of our continuing vested interest, but also we have violated a divine charge. If we treat them well, not only is our tangible reward on Earth immediate and bountiful, but also we may expect the subsequent benefits due to loyal and faithful servants. The consumption of nature thus becomes a *cause*.

Now, whether many of our pioneer forefathers on this continent did or did not act in daily dread of this particular interpretation of the scriptures I do not know. I doubt it; I think the subdual style — conquer in the name of the Lord — is more appropriate to the Calvinist persuasion. I know that from their day to ours there has been an aura of godliness about the sensitive treatment of certain natural phenomena, most particularly soils; however, sanctimoniousness in wildlife conservation affairs has not usually been notable. The "sacred trust" notion has not moved much beyond agriculture. There has, however, been an interesting recent increase in participation on the part of theologians writing on broader matters of environmental concern that tends to echo the "stewardship" theme.

There is another important element. If service to the greater system (the biological continuity) is blessed, so too must be service to our descendants. Stewardship is most easily understood and accepted, I think, when suggested in the context of our children, and theirs. Entreaties on behalf of my offspring are more likely to commend my interest than those more nebulously stated. I am even moved, perhaps, to respond to entreaties on behalf of *your* children. The obligation to future generations is one few would disavow. It is a compelling appeal – to the point of occasional triteness.

Although we hear of this obligation too often from the lips of politicians and we read it too frequently on the letterheads of private organizations, it is rare for anyone to fly in the face of it. If motherhood is no longer quite so sacred as it once was, our duty to those human beings now in place and to be anticipated in future is inescapable. If you won't do it for yourself, or even for the greater biological or theological purpose, you will do it for future generations. You have the right to use nature in your own time, and you have the duty to pass it on in at least as good shape as it was when you received it – preferably better. Thus, the good steward of natural resources, including wildlife, is enjoined by God, by contemporary conservationists, and by those not yet born. This is not a threat of reprisal. It is a matter of conscience.

Here, still in the rural mode, the self-interest of the good husbandman is again the major theme. The appeal is more direct and perhaps more effective because it involves a shorter time frame than the stewardship notion, offering a more immediately foreseeable dividend than mere husbandry for its own and your future sake. The concept of harvest is one everyone can grasp, or could grasp in the days before urban concentration, when the figure of speech was first used with respect to wildlife.

Although its origin is obviously agricultural, the harvest metaphor was addressed most particularly to those who kill either for recreation or for profit. It is chiefly they and their advocates, agents, and creatures in government service who continue to use it today. The rationale is quite simple. Most plants and animals produce in a given year more offspring than their environment can support. The codfish and the lobster are well-known examples. In nature, this seems to be a form of life insurance. If some unusual or catastrophic set-back were to strike the breeding population, there is an available pool from which fast recruitment may come. In most years this reserve is not required, and it is resorbed by the system. It is recycled. In the case of game species, this is termed the "harvestable surplus." Wise use of the resource indicates that this "surplus" may be taken without affecting capital. Each breeding season, then, produces such a dividend to be "harvested" for fun and profit. Otherwise, it would be "wasted."

"Harvest" is – or was, in earlier times – a compelling thought in North America. There was much traditional joyful imagery attached to it: bounty, togetherness, colour, turkeys, cider, pumpkins, wheat sheaves, hay rides, and all the rest of it. Bringing in the cornucopia. We were raised on it, and most of us have warm, positive cultural images of harvesting. Even if it is little more than folklore it is appealing. (As in most orgies of nostalgia, we forget about the work that was involved in the good old days.)

It is interesting that in Britain, which did not experience the North American variant of Protestantism, and where space dictates that agriculture is quite different, the euphemism for sport killing is "culling." Culling is a gardener's word, not a farmer's. It is the selection of individual plants in a bed, not the wholesale reaping of them. Presumably the British figure of speech is just as useful to the "sporting" community as our own.

The fact that the "harvest" metaphor is still everyday talk in North America, especially among those concerned with species that are killed for human recreation, is testimony to its effectiveness when you remember that in the biosphere there is no harvestable surplus of anything. Waste is unknown. All of that reservoir of organic material that is not brought into breeding populations is fed back into the system. That "excess" *is* the support system. It is food, chemical compounds, nutrients. It fuels the very processes that sustain breeding populations and give rise to next year's.

Significantly, the wolf that takes the caribou or the falcon that takes the teal or the whale that grazes the krill is never seen to be "harvesting." Presumably that is because that individual consumer simply takes what comes its way and has no conscious investment in it. The meaning of "harvest" is that someone worked, usually very hard, for it. Someone planted and cultivated and tended and hoed and watered and raised it. These the wolf and the falcon and the whale clearly did not do. That the "sport" shooter did no more is also clear.

Profound implications regarding the relationship of man and wildlife notwithstanding, the "harvest" argument works, and has worked for several generations. That it continues to work so effectively today has certain other significance, to which I shall return.

Essentially, the notion of "harvesting" is a logical extension of the "husbandry" argument in a way that has much more immediacy and apparently wider appeal than mere caretaking.

Future Resources

Here we have a further development of both husbandry and harvest themes. Although cast in the future, it must derive its substance from the past. Earlier generations of harvesters unwisely consumed all of the passenger pigeons, great auks, Labrador ducks, and Carolina parakeets, and most of the Eskimo curlews, bison, pronghorns, sea otters, grey whales, and many more. Their treatment of those "resources" was unintelligent by today's standards and we have been the poorer for it. The fact that this generation is in the process of using up most of the Atlantic salmon, African elephants, and all of the blue whales, will no doubt be described as unwise and unintelligent by the next.

This argument, however, was not originally and is not now restricted to wildlife. It applies across the "resource" spectrum and, indeed, may find application with some of the so-called "non-renewables," to which most of the notions of conservation as a dynamic activity seem inappropriate. A shake-up of the plastics industry, for example, might at least prolong if not guarantee the supply of crude oil and other organic substances for more appropriate purposes.

As it applies to wildlife, this injunction is not much different from its fellows in the "wise use" category. It underlines utility; it perceives resources as exclusively human property; it tries to emphasize good husbandry and common sense. It is sane, sound, and something short of inspirational. It assumes self-interest, and that we have both the capacity and the will to extend our self-interest beyond our individual lives. It is not what the young would call a "grabber."

Science

This aspect of the "self-interested wise use" argument is much more recent than all of the foregoing. It arose later, and is thus more "relevant"; its appeal is strong in most quarters today. I remember when I first encountered it, many years ago, I thought that at last here was ammunition that seemed very promising indeed. Used judiciously, it would help to transcend the rural folksy mode in which (quite properly, in its time) conservation had been cast; it would appeal to those who simply were not interested in compost heaps and water-sheds and tree farming and topsoils and trout hatcheries, but who *were* interested in the advancement of human power in the universe through science. In other words, here appeared to be the beginning of an argument that just might allow wildlife preservation to penetrate the meticulously ordered anthropocentric world of high civilization. Wildlife as the precious raw material of sacred science! It gave me a marvellous morale-boosting lift at the time.

The argument is logically straightforward. If you lose something before you have a chance to study and understand it – even an animal or plant species – then not only you but all of science (human knowledge) will be the poorer. This is where the judicious use comes in. You do not publicly advocate preserving the hairy-nosed wombat for the sake of science, because nobody will take you seriously. But you do advocate the preservation of tropical forests, which contain plants of potentially inestimable worth as, perhaps, poisons or cosmetics. If you let these forests go, who knows what might be lost to mankind, forever? If you lose the sea lion or the fur seal before you understand the biochemical processes of delayed implantation (of the fertilized egg in the uterus), who can tell what alleviation or indeed enhancement of the human condition will have been denied to us? And so on.

This appeal — and it is a reasonably persuasive one even in circles where the notion of wildlife preservation is without meaning otherwise — is of course packaged in utilitarian terms. Although the argument is made in terms of pure science, and is thus of the highest respectability, the real message is to application — technique. This is where much of its strength lies. Even the most impatient and headstrong among the technocrats acknowledges that technique arises out of "pure" science, and that in the final analysis man's mastery of the observable world depends on man's ultimate knowledge of that world. Thus is the appeal for conservation made respectable in contemporary terminology. The benefit to wildlife is slim.

Very recently, this argument has been extended in ways that were undreamed-of in my earlier experience. Ecology, the study of the process inter-relationships of living beings in their environments, has become one of the most popular of sciences. It exists also as an "applied" science. The "wise use for the good of science" argument has allowed many varieties of ecologic research to develop, on the grounds that if you do not know how the biophysical system works, then your chances of manipulating that system in the advancement of human affairs are limited. This has much strengthened the impetus to preserve natural areas, often as representative examples of various habitat types. The appeal is made in entirely utilitarian terms, but there are incidental benefits to both plant-animal associations and to individual species.

Pure knowledge has been the sometimes unintended beneficiary of the relatively new activity that has come to be known as "environmental impact assessment." EIA, as it is

known among its practitioners, is interesting and enlightening. When some grandly glittering construction scheme is
proposed, appropriate governmental authority may require
that a token investigation be made of its possible social, economic, biologic, or ecologic consequences. (The political
results will have been weighed well before the plan is made
public.) In the process of such investigation, information
may be uncovered that may be of no interest whatever to the
proponents of the project, nor indeed to the bureaucrats, but
that may add to the pool of information. Thus we have gained
knowledge, it must be said, that we would never have gained
without the funding that surrounds large building projects.
Wise use – here meaning ecologic field studies – can indeed
mean the advancement of science.

The quality of science so delivered, however, may be
variable. Bad science as well as good science emerges. Many
people have observed that the slapdash way in which so much
of this slickly packaged work is done and the casual level of
review to which it is subjected are open invitations to disaster, both for science and for wildlife.

Wildlife preservation can never be an effective argument for the undertaking of these sorts of studies. Since the
reasons for the studies themselves are entirely political
(strategic, or in many cases, merely tactical), the guardian-
ship of wildlife populations, species, or associations is
never a factor in a motivational sense. But, like the good
Presbyterians we conservationists are, we long ago learned
to accept the collection plate as it comes, and not to ask too
many questions.

EIA is a grandiloquent fraud, a hoax, and a con. Others
have seen it as both a boondoggle and a subterfuge. It is one
of the deceitful co-options of the concepts of ecology and
environment of which I spoke earlier. While sanctimoniously

reciting the catechism of "environmentalism," it anoints and blesses the process of "development," takes the initiative from the preservationists, and, in most cases, effectively bulldozes, gravels, and hardtops the road for the technomachine. Ecology is thus used as a tool to permit "developers" to continue to do what they have always done. The only difference is that "environmental impact is to be minimized to an acceptable level." What is minimal impact? What is acceptable impact? Acceptable to whom? Wildlife, alas, cannot be interviewed.

EIA may not be good science, and it may not be conservation, but it is excellent business. Like its cousins, the regional master plan and the evaluation of land capability, it does contrive to circulate much money, the tab for which you and I, and the silent majority that is wildlife, will ultimately pay. Also, it tends to shore up the ancient human hallucination having to do with the predictability and thus the controllability of all things.

The scientific argument for wildlife conservation has come to be widely used in a somewhat unexpected quarter – zoos. Many modern zoos have made much of the need to protect (and of course to display) critically rare or endangered species, as a contribution to the larger conservation movement. Historically, as the supply of new and replacement (the turnover in some of the newest establishments is appalling) animals from the wild began to dry up dramatically, zoos were forced to develop captive breeding programs of their own, for a relatively few species. This had immediate results in veterinary medicine, among other applied sciences. It also had "box-office" value arising from the publicity surrounding a list of endangered species such as the great apes, some antelopes, and others. Such publicity is generated (albeit in quite different forms) both by conservation organizations and by the zoos themselves. The benefit to the zoo is obvious; the benefit to wildlife conservation, so long as the animals remain captive and are not released to the wild, is zero. Indeed, natural habitat destruction may proceed at an accelerated rate now that the public has been assured that the species is "safe." The benefit to some aspects of scientific and technical knowledge is said to be considerable.

Here we have again the utilitarian theme (research, education), although in this case pure knowledge is most frequently identified as the contribution of zoos. Also, of course, we have wildlife as "resource" in the usual sense of human proprietorship – in this case for captive display in the interest of entertainment. Again, the fact that we have neglected to interview the animals by way of opinion survey is somehow overlooked.

"Wise Use" in Summary

"Wise use" is the hard, germinal kernel of the self-interest approach to wildlife conservation argument. Wildlife is a renewable resource. Treated wisely and intelligently, it can benefit today's users and those of future generations. Your right of access to the resource is accompanied by a duty, not to the resource, but to today's and tomorrow's publics. The resource is yours to enjoy and to protect. The careful use of the resource is not only prudent but also, in a human ethical sense, obligatory.

Throughout, wildlife is seen solely as a human asset. There is not the slightest question of any intrinsic interest being vested in wildlife itself. It is a human commodity.

THE "QUALITY OF LIFE" ARGUMENT

This approach has emerged more recently as a popular argument for wildlife preservation, although the poets have expressed it metaphorically for centuries. Still very near to the core of the self-interest sphere, it tends to take as given most of the "wise use" implications (husbandry, responsibility, future generations), but it concentrates on a qualitative as opposed to a measurable payoff. In general, this argument has enjoyed its greatest effectiveness and success in realms that are very remote from the wildlife question – toxic pollution, urban horrors, the population blight, industrial indecencies, and recreational demands.

This one meshes well (indeed much better than "wise use" can do) with the conventional picture of a civilized society. Life quality as a notion rests well with many branches of our society that are relatively untouched by rural and wildland imagery. Because the overwhelming majority of us are city-dwellers and, as such, are especially sensitive to the comforts and discomforts of an entirely fabricated environment, the "quality of life" argument may be easily carried in terms that everyone understands – physical amenities, arts and letters, social relationships – the humanistic tradition generally.

What does this have to do with preservation of wildlife? At first glance, very little. But there are ways. For one thing, the protection of nature is something that obviously befits the sensitive and genteel human being. Because many perceptions of life quality have strong narcissistic overtones, it suits the individual and collective self-image to patronize "those who cannot speak for themselves" in much the same way as we would patronize the arts and sciences. Or the poor, needy, and afflicted. It is the proper thing to do. This can be used – not as an argument, of course, because it is not one – but it can be used as a device.

Among conservationists the quality of life concept is very often expressed in terms of *environmental* quality. This helps to narrow the focus slightly, especially if we are then able to suspend the wider sociocultural aspects of "environment" to concentrate on the biophysical. But even here there is a problem. Virtually everyone seems to agree on "environmental quality" as a valid policy goal, but I have not yet seen a firm definition of what it means. This is an important hindrance.

You and I may agree on many things, but we will not agree on all things. Our appreciations of environmental quality may be absurdly different. The arctic is of very high environmental quality for seals, but of very low quality for flamingoes. The contemporary urban environment is of very high quality for rats and roaches, but of very low quality for moose and for millions of people.

You might be tempted to take issue with this, on the grounds that the notion of "quality" is entirely inappropriate and indeed irrelevant to an animal's ability to survive in a given natural situation. Of course a flamingo cannot survive in the arctic, you say. Either an animal is adapted to a given environment or it is not; the question has nothing to do with "quality." In response to this, I would thereupon leaf through my field guide and identify you as a species of 'ologist, seeing all things through reductionist human eyes, and denying the capacity for qualitative experience in all beings except human beings. I would accuse you of taking craven refuge in "objectivity," which in my book is listed under cop-outs. I choose to believe that nonhuman beings *do* perceive quality and its absence in their environments, and that they behave accordingly. Watch a wolf in a cage; watch a porpoise in the open sea.

Even if we were to adopt a completely self-interested (anthropocentric, as opposed to biocentric) definition of environmental quality, your quality would almost certainly not be mine, nor mine yours. I think that both of us, however, would feel that environmental quality involves at least some element of choice, something having to do with keeping the options open. (This implies that we are able to recognize options for what they are – possibly a foolhardy assumption.)

To the naturalist, environmental quality means, among other things, heterogeneity. It means variety, diversity, options. Homogeneity (uninterrupted wheat fields, tree farms, cities, highways, feedlots) is a kind of ecologic heresy. The concept of a healthy ecosystem is one with many parts, many roles, many inter-relationships, many alternatives, many structures and functions, many ways of doing things. It is dynamic and it flows and changes. The resilient ecosystem provides options for its parts and processes in the event of disturbance. It has escape valves. The monoculture, on the other hand – the pure culture of Scots pine, rice, or people – is an open invitation to disease, parasitism, and destruction. The reduction of parts and processes, roles and alternatives, can weaken a system and make it vulnerable. The same is true, obviously, of a society or a culture.

Life quality means life options, options for change. To the student of evolutionary biology, every variation in every individual organism is an option for change. It may not be "used" immediately – or ever – but it is there, against possibilities. Evolution is opportunistic: new variations, each of which is an individual "experiment," continually probe and test the changing environment for new opportunities. When the option presents itself, a new variant may become a new ecotype, a new race, perhaps even a new species. There are variations in anatomy, in physiology, and in behaviour. These variations evolve, in concert with changing environments, over time.

There are variations also in human cultures, beliefs, values, traditions, assumptions, even perceptions. Or rather there used to be. The culture of the western techno-mechanistic (we sometimes call it "developed!") world seems to me to be pathetically homogeneous. A colossal irony is that we apparently perceive this homogeneity (one world – in our image) to be good. On the other hand, the "quality of life" argument implies that there may be options.

I readily grant that the connection with wildlife preservation is getting tenuous here, except for those who see wildlife as a desirable option, the existence of which can contribute to their particular model of life quality. This has never been a strong contender in an operational sense. Wildlife can be an unintended beneficiary, however, as, for example, when pieces of attractive landscape are set aside for scenic or recreational purposes. Here, those forms of animal and plant life that are able to tolerate whatever intensity of human use develops, persist. And of course there are those animals desired for the killing which are carefully "husbanded" as part of the life quality of their peculiar human constituency.

Here I would tentatively venture to explore a more expanded notion of "life quality" which is not yet a stock argument. This is the biocentric approach. It is advocated by numerous individuals but not, I fear, sufficiently widely or intelligibly to be of much significance yet. (This is not to suggest that I disagree with it – on the contrary – merely to say that at the moment it has gone exactly nowhere.) In this approach, "quality" is simply the phenomenon of living, of which man is a part, but only a part. The quality lies in the richness of the life experience *within* flows and forms and processes that are not of human manufacture and that function as a total and heuristic happening. The experience lies in being in and part of it. As the richness and variety of the environmental whole are reduced, simplified, and degraded by human activity, so the richness and variety of the individual human life experience are proportionately affected. In this case "quality" becomes the maintenance of the life complex in all of its manifestations, allowing the process to exercise its own options, as it always did in prehuman times. All this in the interest of fuller human experience.

This starts to drift toward the ethical argument, which will be discussed later. Since we are still, at this stage, in the realm of human self-interest, perhaps I should not bring up ethics at all. However, ethics, whether classical or not, seem to carry with them at least the promise if not the presence of reciprocity. It might be that the fullest human experience can only be realized in an ethical approach to the living surround; that it would be in our own interest. More of this in a more appropriate place.

This is the right place, however, for a look at the *aesthetic* argument for life quality as it contributes to the human self-interest. The benefit to the individual beholder is vivid and present. There are many compelling arguments for wildlife framed only in terms of beauty. Much of this is delivered to us by way of graphic art, including photography. (This is not to deny the long tradition of "nature writing," of which I shall have something to say later; I am thinking of our special primate fondness for visual immediacy.)

Much if not most of the conventional aesthetic argument for wildlife is presented not only in visual terms but also in environmental context. It is easier to present the aesthetic appeal of a walrus in the dazzling shimmer of silver and blue and green that surrounds it, than in silhouette. It is the frozen lake and its looming rim of spruce that make the wolf "real." Even the bird-of-paradise looks better against its deep and lightless jungle backdrop than in vignette. What is really "sold" in pictorial wildlife aesthetics is the natural environment, not its pieces. For its part, the bird-of-paradise "completes" the forest scene, makes it whole. I suspect most people, when admiring the appearance of a leopard in a zoo, project it back into its native forest. Many people feel sad at a zoo. That too is an aesthetic experience.

The beauty of aesthetics is that in spite of our infatuation with opinion surveys it cannot be quantified. Aesthetics does not rely for its respectability on supporting hard data. (In view of this, one occasionally wonders how it is that in the present time aesthetics contrives to survive at all!) Apart from its original meaning as a branch of philosophy having to do with "the beautiful," aesthetics is largely a matter of individual human perception. To this extent the value of aesthetics is heightened – or lessened – according to one's individual perceptual predisposition.

In classical aesthetics, the beauty of a thing is weighed by comparison with its "ideal archetype" – its prototype or model – itself often a stereotype, or what we might call an institutionalized image. It is the product of cultural convention. It changes over time, but slowly. Hence the difficulty, it would appear, experienced by the creator of atonal music or cubist portraiture or pointillist landscape or geodesic domes in cracking the vested resistance to change. Each of us varies from the cultural norm in these things, but not much. Even though aesthetics is very much in the eye, ear, and mind of the individual experiencer, that experiencer has been well and truly conditioned by the traditional cultural environment. The possibility of aesthetics being much more than pure "nurture" seems to me very remote. (No doubt the aesthetic *capacity* is inherent in us; the content is inserted.) Every creative artist knows how calcified, rigid, and unyielding culturally institutionalized images of "the beautiful" have a way of becoming.

At the present time, the (visual) aesthetic appeal for wildlife is packaged almost entirely in conventional landscape. We are receptive to it because it is in the cultural background. Traditional landscape aesthetics, of course, has nothing whatever to do with wildlife, but it is often influenced by the presence or absence of man. A somewhat dismaying (because it is probably accurate) formula has been put forward by Nan Fairbrother: landscape = habitat + man. It is, as she says, an unstable equation because it is forever changing – changed by a creature who is himself changing. Landscape, she points out, is not a static background, but the interaction of a society and the habitat it lives in. All is change.

As a naturalist I am very much aware of the process of change, and so I feel that in that sense Fairbrother is absolutely correct. Nothing can be static; even institutional-ized images can change over time. What *is* however alarmingly close to static is the perceived central role of man in all things, even landscape. That particular hallucination is firm.

From the traditional point of view we hear a great deal more about "man in the landscape" than we do about simple "landscape." To the committed humanist observer, a land-scape may be "empty" if man or the evidence of man is missing. At the very least he will want to see the *potential* for man in that landscape. Thus the arctic tundra is a wasteland, the moon barren and empty. A landscape may only be "com-plete" if man or the evidence of man or the potential for man is somewhere visible or capable of being inferred. To the wilderness "purist," on the other hand, man (except in very low numbers and of course except for himself) may not be appropriate at all.

Usually we seem to discover the "man in the landscape" aesthetic at its most mature development, where man has been in residence for hundreds or thousands of years, to a point at which human evidence, in the appreciative observer's eye, is part of the landform itself. A point where man is "in harmony" with landscape. Spain comes to mind here. But we are never told what "harmony" means. If it means the man-induced

Mediterranean deserts or the horizon-to-horizon rice paddies of the far east, then the euphony escapes me, because there nature has been wholly subdued and nowhere shows her face. Wildlife has long since been almost entirely eradicated. As far as human settlements are concerned, "man in harmony with nature" is meaningless gibberish, despite the facility and frequency with which we intone it.

All the same, we like to keep *technical* man's works in perspective. On a Precambrian lake, a canoe "belongs"; a motorboat does not. On the same lake in winter, a dogsled is appropriate; a snowmobile is not. In a forest landscape, the loggers should be using horses, not tractors. I am not going to pursue this further, except to suggest that there is a very long and deep romantic tradition involved here – very Anglo, I think – and it is just as misleading as most other frontier folklore.

Generally, wildlife species become the incidental – almost accidental – beneficiaries of the appreciation of the aesthetic properties of their immediate habitats. They benefit not because there is an ideal archetype of the living being itself. Admittedly some animals (male deer, Canada geese, mallards) and white pines qualify. But they are overwhelmed by cattle, wheat, grain elevators, sailboats, cottages, decoys, kayaks, and people. Here, to the hairy-nosed wombat problem I would add the stinkpot turtle problem. Most wildlife species are not represented by ideal archetypes in our cultural catalogue.

In spite of this, for generations there have been artists who specialized in animals, and whose work has ranged from the most highly technical scientific illustration to the full-fledged elegance of creative portraiture, even impressionism. Several superb practitioners are active today. I feel, however, that in the world of art appreciation their recognition has come more from their talents than from their subject matter. They paint wildlife because they love wildlife, but they also happen to be artists of the first rank. In the salons, the content may well be incidental. I have found no sparking synapse between animal art and operational wildlife preservation.

Within limits, however, the aesthetic argument for wildlife is good. It is attractive to the self-interest in the sense that it adds new or reveals latent options for "life quality." It is most effective, one would suppose, with those who have had prior experience of the phenomenon pictured, either directly or indirectly.

It always seems a bit strange that, when reviewing our sundry aesthetic models, there does not seem to have developed any culturally accepted archetype of the beautiful in terms of such non-visual abstracts as *process*. The nearest thing we seem to have is the appreciation of form in music or poetry or dance – form, as opposed to specific content. This, as a kind of process, we understand and appreciate aesthetically. But we have not developed an aesthetic of *life* process. That is because our culture is essentially abiotic.

The life system – the process – is beautiful. Anything that tends to simplify or obstruct or alter or damage the process is unbeautiful. A wounded, limping life process, like a rotting city core, is ugly. That which is ugly may quite properly be described as disadvantageous to the system. (We may be either self-interestedly anthropocentric here, or biocentric, when we speak of the "support" system.) This is not merely a value statement; it is a factual statement. We are surrounded by the evidence of it. We live in the evidence of it.

Some years ago I chose as my archetype of the *un*beautiful the 45-gallon fuel drum. On the arctic tundra it not only looks like a suppurating chancre on the cheek of Mona Lisa, but also its implications for wildlife everywhere in the world are hideous. If you want a symbol, there it is. I find it more informative than a cute cartoon of a giant panda. There are many other such examples, according to your individual experience, perspective, and predilection.

In sum, the "quality of life" argument for the preservation of wildlife stresses human options, chiefly experiential, and offers a variety of plants, animals, and environments as rich potential fare in the individual and collective interest. A future orientation is implicit if not always stated. It is very weak on life process. The use is "nonconsumptive," but the notion of proprietorship remains firm. The flow of benefits continues to be one way.

THE "ECOCATASTROPHE" (DOOMSDAY) ARGUMENT

This is the scare strategy – the most brutally direct of the self-interest approaches to conservation. It glosses over nothing, etching possible futures in stark and bold relief. There is no waffling, no circling, no obliqueness, little euphemism. This, ladies and gentlemen, is the way it is – or the way it is going to be – unless . . . Heavy handed, perhaps, but refreshingly straightforward and free of rationalization. An apparent weakness is that it assumes that people will listen – and act.

The "gloom and doom" formula, which all of us have toyed with at one time or another, is too familiar to require more than brief mention. In a very real way it reminds me of the evangelical technique of earlier times: scare the living bejesus out of them; then, with exquisite timing, flourish the recipe for salvation. For maximum effectiveness it depends on equal measures of belief and doubt on the part of the audience; hard evidence is secondary.

With or without a litany of facts, the salvationist works on uncertainty, and sells immortality insurance on that basis. But of course what the fire-and-brimstone preacher is digging at is *individual* fear, *personal* fear. I know of no way in which the tribulations of wildlife can scare anyone. The fact that evidence of *human* peril surrounds us (indeed the evidence fairly overwhelms those who understand what they see in front of them) should make the delivery of his payload somewhat easier for the ecocatastrophist than for the salvationist. But it does not. Billy Graham draws bigger crowds than Paul Ehrlich. Belief is a question we shall get into later.

The doomsday argument should be potent. There are problems the magnitude of which we can barely envisage at this stage, and worse is yet to come, unless . . . This is surely the ultimate appeal to the self-interest. So far as I can determine, it has absolutely no relevance to wildlife preservation. When human survival is at last on the line, the priorities will become strikingly clear. Wildlife could of course emerge as a second-generation beneficiary should we choose to listen to the ecocatastrophists and take evasive action in the interest of our own skins. This is not, however, part of the argument, either for wildlife or for the self-interest.

ABOUT SELF-INTEREST

Before we go further, it will be useful to sum up those arguments for conservation that are based in individual and collective human self-interest, as put forward here. The most fundamental message is: *if we can't be good, at least we can be prudent.* The message has been delivered historically and is delivered today in a number of ways: the "wise use" arguments involve husbandry, stewardship, harvest, future resources, science, and technology; the "quality of life" argument stresses present and future options and aesthetics; the "eco-catastrophe" or "doomsday" argument is essentially "act now or pay later."

The guts of the self-interest family of arguments is that they are entirely and exclusively man-oriented, anthropocentric. Whether it is directed to individual, group, nation, or species, the appeal is to the human being and the human interest. The future may also be presented in the form of subsequent human generations. We may frame the argument either in the investment-dividend or the waste-bankruptcy mode; the metaphor may vary, but the message is the same.

Throughout, we assume nature as "resource," whether for physical use or as a source of aesthetic enjoyment. In this sense, living sensate wildlife beings are no different from water, soils, and landforms, all of which were set in place by a beneficent nature expressly for human purposes. Whether man is good steward or renegade, whether answerable to God or to the biosystem or to future human generations or not, there is no question about the locus of vested power and authority on Earth. This is illustrated best, I think, in the monumentally dull-witted arrogance of the concept of "harvest" as applied to wildlife species.

In almost all cases, save commercially or technically important species, which are commodities, the role of wildlife is that of incidental or spin-off beneficiary from "resource management" policies and practices. Even in aesthetics (endangered species excepted), wildlife benefits as landscapes and wildlands benefit, rarely for its own sake.

For its logical integrity, the self-interest argument *must* see wildlife as a "resource," a commodity, a utility. This brings us back to the distinction between "renewable" and "non-renewable" resources. I no longer believe that there is, in practice, such a thing as a "renewable" resource. Once a thing is perceived as having some utility – any utility – and is thus perceived as a "resource," its depletion is only a matter of time. I know of no wildlife that is being "renewed" anywhere – not yellow birch or hemlock or anchovies or marlins or leopards or salmon or bowhead whales or anything else. "Renewable resource" is self-contradictory incoherence, at least as applied to wildlife.

If "resource" continues to mean something that is put to human use, then no resource is renewable. Our demands have quite outstripped the capacity of those resources to satisfy them, much less to satisfy them on a "sustainable" basis. And we are, of course, never satisfied. Our insatiability is a vast subject and one I cannot get into here, except to remind you that there is no scrap of evidence, anywhere, to indicate that demands can be reduced. All of our efforts are directed toward increasing *supply*. Almost all, that is; the "sporting" and recreational industries do put considerable effort into beefing up demand. The best we seem to be able to do in the way of "renewing" something is to stop consuming it for a while and hope that nature will do the rest.

In our society, if a thing has value then it *must be used*. Otherwise we could not put a value to it. The word "value" as applied to wildlife derives from the *utilitarian imperative*, which is a part of our imperialistic stance toward nature as a whole. This underlies even the scientific-technical argument for wildlife preservation: who knows what uses might be discovered one day? Ironically enough, it is also implicit in the familiar "*No use* is a legitimate *use*." Now this one is very glib and very fine on the surface, but it freely admits that conservation must present *all* of its case in the utilitarian mode. This is more than saddening; it is frightening. Conservation, far from confronting the utilitarian imperative, chooses to legitimize it and argue from it. This, for our movement, I perceive as a species of death wish.

With very precise (commercial) exceptions, the self-interest argument has not and cannot preserve wildlife. It can perpetuate some specifically desired or most-favoured forms, no more. The preservation of wildlife *for its own sake*, with no implication of use, is antithetical to the self-interest position. By its very emphasis on the utilitarian imperative, the latter drives the conceptual wedge between man and nature ever deeper, thus reducing the possibility of the achievement of wildlife preservation in the ideal sense. This is not a comforting thought, for many conservationists would claim that self-interest is the only argument we have.

The only wildlife that can be (conventionally) *proven* to be in the human self-interest are commercial species that we can describe in dollar terms. Dollar values are attached to individuals of certain endangered species in the zoo trade also. Beyond this we have the very nebulous category of those wildlife forms desired for aesthetic (quality) reasons, which in most instances centre on landscapes and recreational possibilities rather than individual species.

The overwhelming mass of wildlife species are not at present commercially desirable, and although the list of rare and threatened species is depressingly long, it is not yet great relative to the absolute number of plant and animal species that are still with us. But it grows rapidly. It is my waking nightmare to think that a species must be either commercially valuable or endangered in order to be seriously considered for active protection. But if one persists in seeing wildlife as a "resource," and thus in accepting the burden of demonstrating its utility in the human interest, this is where the argument must necessarily and inevitably lead. There is no escape.

To this point, apart from a brief note, we have not looked at those uses of wildlife usually termed "non-consumptive." For many years wildlife conservationists, for purposes of argument and explication, drew a distinction between those activities that actually "consumed" wildlife and those that did not. The distinction was useful at one time because it tended to underscore the need for different approaches to wildlife "management" under different circumstances and conditions, and especially to emphasize the need for protection of non-commercial species in the interest of a body of people who wanted wildlife for "passive" recreational purposes, no more. Simply for the viewing.

This was fine just as long as there were not too many people who wanted wildlife simply for the viewing. It will not be necessary here to go into all the recent revelations about the physical impact of the tourist stampede on parks, sanctuaries, and game reserves all over the world. Recreational ("non-consumptive") uses have become an enormous problem. Indeed there is no such animal as the "non-consumptive" user of wildlife. We are having to "retire" campsites even in Canada, where there is alleged to be infinite room. To my certain knowledge, a succession of zebra-striped vans with their loads of instamatic-bearing tourists is infinitely more damaging to the ground cover and thus to the *total* biosystem in African parks than any wretched psychotic in his Abercrombie and Fitch hunting costume can possibly be. This we know. The volume and the impact of tourism in all parts of the world are absolutely staggering.

Just as I can no longer see wildlife as a "renewable resource," I can no longer think of a "non-consumptive" use of wildlife. If it is a use, it consumes. The terms are archaic and no longer have meaning; worse, the arguments they imply are destructive of the goal itself. By suggesting that wildlife is "renewable" you sanctify the notion of "management," you confirm and ratify the assumed role of wildlife in the service of man, and you also imply that it *is* in fact being renewed. Nothing could be more misleading, or fallacious.

By suggesting that recreational uses of wildlife other than outright killing are "non-consumptive" you overlook and thus worsen recreational impact, and once again you legitimize the utilitarian imperative.

It seems that such notions as renewability and non-consumptiveness actually contradict the real goals of wildlife preservation. The standard self-interest argument, even at the aesthetic level, acknowledges no earthly role or place for wildlife beyond the human interest, no existence beyond the human purpose. If not overtly, certainly by implication, it nurtures a perception of wildlife that is radically at odds with preservation. The very attitudes the argument perpetuates have always been our chief obstacle. This is the "loop" into which we have fallen.

Most important, in a practical sense, the utilitarian argument leaves in limbo the huge majority of wildlife species for which no such value can be demonstrated. Dependence on the self-interest argument has knotted conservation in an impossible double bind, our stated case actually cementing us into our most basic problems. Such perhaps was not always the case, but it is today. To admit wildlife into the human cosmos solely as a commodity in the human interest is to make preservation an absolute impossibility.

Finally, self-interest is a risky approach in any case. It is well-known that practices in the self-interest have a way of being abandoned and forgotten if the predicted difficulty does not loom over the horizon soon enough to keep them alive. If the "danger" cannot be shown to be actual and immediate (and who, after all, would feel threatened by the extinction of the Cambodian kouprey or the New Caledonian kagu?), self-interest tends to evaporate as any kind of motive. The trumpets are not sounding the call to judgement just yet. We still have time.

The argument of self-interest is fundamentally contradictory to the wildlife preservation purpose, and in actuality works against that purpose. Its long tradition is one of the prime reasons for our failure.

ETHICS

Rightly or wrongly, I have arbitrarily separated this examination of the ethical argument for wildlife from the self-interest section, to concentrate on pure altruism. Current "sociobiology" maintains that there is a deep and abiding vested genetic interest in nonhuman altruistic behaviour, but I choose to set that to one side for present purposes. Whether it is legitimate to consider ethics outside the realm of self-interest is open to question, perhaps, but I shall proceed regardless.

There has been much call for an "environmental ethic," and many good minds have spent much time in trying to come to grips with it. Most of this work has centred around the human (especially urban) environment, in much the same way as "quality of life" explorations have done. An ethical approach specifically in terms of wildlife is rare. Although an undefined "ethos" would seem to be the prime motivation of many of us, to the present time it has defied institutionalization among our organizations. This is not especially difficult to understand when we acknowledge that up until now ethics have been concerned almost exclusively with human individual-to-individual and individual-to-society relationships.

Although we do use the term "ethics" broadly and I think somewhat sloppily in our arguments on behalf of wildlife, we come back to the "quality of life" problem in that no one has yet advanced an acceptable (meaning workable) definition of what a wildlife ethic might look like. Ethics, in the non-academic sense, are simply principles of good or moral behaviour. To urge a "wildlife ethic" is merely an injunction to behave morally or decently toward wildlife. This would have a variety of possible interpretations through the perceptual spectrum of the sundry vested human interests in wildlife. Also, it does not help us much in the absence of background on what sort of behaviour might be considered good, moral, or proper, from the point of view of wildlife. We cannot ask them. The Golden Rule of a Tasmanian devil might be at odds with that of a planarian, and there is some possibility that even the human mind might not be able to comprehend either.

Let us begin with some of the more conventional arguments, which may help in the search for implications and refinements.

OUR MORAL DUTY TO FUTURE GENERATIONS

When viewed in the ethical frame, this has slightly different meaning than it did in the self-interest context. We have discussed the notion of "stewardship" on behalf of our descendants — the obligation to hand over that which has been in our temporary custodianship in as good shape as we can manage. This is an entirely anthropocentric appeal, and an apparently useful one. If we try to move it somewhat beyond tangible self-interest (the bird actually in the hand), it becomes somewhat more appealing.

In a way, this becomes a kind of "quality of life" stewardship — a custodianship — not of material assets in the form of commodities, but of a set of choices or options. We have a moral duty not to foreclose options on behalf of those who will follow us. There is no shred of reciprocity involved in an ethical arrangement with those who are not going to be around until after our deaths, so in that sense at least this is an altruistic position. There is nothing in it for us; only for them. (That there may be for us a soupçon of private congratulation is forgivable.)

The argument sounds even better when we acknowledge that we haven't the faintest idea, really, how future generations will *perceive* their options. Or indeed how they will perceive "resources." Perhaps they may well see values — or values of different kinds — in items which at the present time we either consider unimportant or fail to see at all. This possibility alone means a moral obligation not to subtract anything now that might be of either material or experiential value to our descendants. Remember — resources *become*.

One example. They tell us that there are so many of the strange boojum (cirio) trees left in Baja California because nobody has thought of any use for them. But the fact that they are "useless" now is no reason to assume that they are going to remain so in perpetuity. So save them; maybe somebody will want them some day, after we have exhausted everything else. There has got to be *some* use for them, or God would not have put them there. This is the utilitarian fundament of the self-interest argument. On the other hand, it might be possible to speculate that boojum trees represent a kind of experiential or qualitative value as yet unrecognized by most people. For certain the world is full of people who have never clapped eyes on one. It would be immoral of us, unethical, to deprive our descendants of that experiential option.

All of it finally hinges on what we consider to be fitting and appropriate options for human posterity. And of course the morality is toward them, not toward wildlife. So it is that we, in our ethical posture, magnanimously bequeath the entire world of nonhuman nature to our heirs and assigns, once again emphasizing the total proprietary interest of man — whether in a quantitative or a qualitative sense — over the "resource." The message is clear and unequivocal. The anthropocentric universe turns, still very much locked in place, no matter how circumspect our behaviour toward our human survivors.

ETHICS AND VALUES

These present an even more difficult tangle of problems — interesting ones. To behave ethically in our society means to behave in conformity with a set of principles having to do with morality and propriety, toward the greater social purpose. Almost everybody knows the rules, and almost everybody behaves accordingly. Breaches of the rules are considered gauche at the very least, criminal at the worst. In general, the system works, because all members of a society are assumed to have access to the ethical code.

However, when we begin to circle around the notion of a "wildlife ethic," we encounter brambles. Let us for the moment see the wildlife ethic as part of a larger, broader "environmental ethic," about which much has been written, more speculated. There is no denying the fact that any ethic, environmental or otherwise, must be expressed in human terms in order to be understood by most of us. Those human terms are of course the human social interest. This of necessity takes us right back to both the "steward-ship" (wise use) and "quality of life" varieties of the self-interest argument. An ethical position having to do with the total environment is construed as an ethical position with relation to those human beings now and in future who must live in that environment. Wildlife, for reasons suggested earlier, is not an important factor here.

So we come to the tough question: what about an ethical position with regard to that part of the total environment that is not human and is not of human manufacture and is not seen to be in the human interest? Is it possible to conceive of an ethic that could be extended outward, as it were, from man to non-man? From human society to nonhuman society, or wild nature? This is a painfully difficult task, even to conceptualize, much less to express in words, still less to implement. Here I am assuming the degree of altruism that would be required to extend unilaterally such an ethic in the certain knowledge that no "normal" reciprocity can be expected.

Historically and traditionally, indeed perhaps ultimately, there is simply no meaning to "ethics" outside the human sphere. This should not come as a surprise to anyone. Even if I believe myself to feel supremely and wholly altruistic in my ethical behaviour toward others, the matter of reciprocity is fundamental. Even for the most saintly among us, there is at least a warm reflected glow. All – or at least a very large majority – of people in a society *must* behave ethically (according to the canons of that society) if the ethics of that society are to mean anything at all. Wild nature, however, does not know our rules and is unable to observe them.

I think that we cannot conceive of a society (a community of self-interest) that could extend beyond our species. Such a society would have an impossible prerequisite – an unequivocal acknowledgements of the *whole* inter-relationship between man and nature. Such an awareness, which would of necessity involve both intellect and intuition, does not exist to any discernible extent in humankind.

I fear, therefore, that when we speak of "environmental ethics" we are very much talking to ourselves. We mean much the same thing as we do in the "quality of life" argument – responsibility toward people. Now, if the self-interest of people indicates that certain policies and practices should be implemented in order to conserve and perpetuate the non-human elements of our environment, then steps may well be taken, but not as an ethical stance toward those elements. It will be done in the self-interest. If the self-interest cannot be demonstrated, you may be certain that western society will develop no ethics with respect to nature.

It is obvious that the issue is essentially cosmological. The concept of the anthropocentric universe dictates everything that follows. As I have pointed out, the absence of an aesthetic for life process is part of the same problem. There is neither an ethical nor an aesthetic possibility for anything but a man-centred and man-ordered cosmos. That is what the technoculture is all about. The problem is metaphysical – not ethical, not aesthetic. The problem is the traditional western humanistic vision of man and his cosmological role.

As you begin to experience the full enormity of this problem, you will begin to fear that much of our effort in the proselytizing of the "wildlife ethic" is quite empty and fruitless. It is not that our audiences disagree with us or resent our argument or are offended by it: it means that *they cannot perceive it*. They literally do not know what we are talking about. To postulate rationally a cultural role for a nonhuman ethic would literally mean turning western metaphysics inside out, and that is a tall order, because it has to do with human purpose and meaning in the universe. I shall return to this.

One way of describing a set of ethical principles is to see it as mirroring the set of values shared by a society. In this sense, ethics and values are pretty much the same thing. We behave ethically toward those things we value; our modes of behaviour spring from those values in the first place; in turn, the ethical behaviour reinforces the values, and so on. This is a very tight lock-step relationship, the parts of which one really cannot separate. For these purposes I am going to pretend that I can, however, in order to look at the question of wildlife values.

There is much talk about the "value" of wildlife, from its aesthetic through recreational and economic to ecological value. These assessments are entirely arbitrary, of course, because they are attached to wildlife unilaterally by ourselves in the absence of dialogue with wildlife. All of us can find niches appropriate to our individual wildlife values. It is not so easy to find places or groupings for our social or cultural wildlife values. One generalization is possible: when we speak of wildlife values we mean values to us, not values to wildlife.

This seems reasonable enough, for it would be difficult to imagine that any species save man is capable of entertaining values as we understand them. A value, of course, does not exist in a vacuum; to be perceived at all it must be in a context. Usually it is a context in which we may apply such abstract (again, arbitrary) measures as good and bad, right and wrong. We like hierarchies: good, better, best. We are compulsive evaluators, in spite of the fact that our valuations have no reality whatever outside our own heads. I doubt that any species save man goes in for this obsessive accounting. I am fairly confident that nonhuman species do not need to entertain such odd concepts because they do not invest much time in rationalizing their appointed places in the perceived hierarchy of the cosmos. They do not seem to require that kind of reassurance.

Since values are abstractions, they have no concrete external reality, and like any other images can be changed at will. What *is* real, I suppose, is the *process* of valuing that goes on inside the individual human skull. I see myself forever entertaining notions that at least to me seem to have merit or usefulness, or that in my own egocentric universe may shed some positive light on me, the unique possessor of those particular notions. Or, I may well share a value – love of birds, let's say – with others, and in the standard pattern of group narcissism that value becomes a means of sorting out our hierarchic place relative to other groups, whom we are then able to perceive as philistines. Remember that "value" implies comparative measurement, whether we acknowledge it or not.

The values attached to wildlife by me or by anybody else have no existence outside our own minds, so we must admit that wildlife, like any other value, has no *intrinsic* worth. It has only the worth that you or I choose to give it. The value is an abstraction, a fabrication, an individual judgement only. Valuing wildlife from a broad social base could only result from large numbers of people sharing both the same perceptions and the same valuing process. Such synchrony is unlikely. The ultimate value, really, is in what the valuing process does for the individual valuer, no more than that. But that of course is everything.

It is interesting that in nature no price and no penalty seem to be attached to the individual existence. (We can and do conceptualize evolutionary and other biological and ecologic values, but these are pure fictions as well, not statements of real worth.) In nature there are no weird and bizarre notions like egalitarianism and the sacred individual to gum up the process. You merely spin out your time – eating, voiding, propagating, and eventually rejoining the system at a simpler level of organization. But since nothing goes away, there is no problem. An existence singularly free of hangups. The problem with valuing is that it attaches universal *meaning* to things.

In all honesty we must acknowledge that only a very modest fraction of people thinks about wildlife *at all*, even fleetingly, much less consciously attaches any value to it, still less ranks that value in any hierarchy of importance (vested interests in commercial species excepted). I am referring here to what we *do*, not what we say. When put to the test, any short-term human value you can name will operationally outweigh any long-term value, to say nothing of any non-human "value" of whatever time dimension.

Ethics and values, though interesting and indeed persuasive in an abstract theoretical sense, have little or no *practical* utility in the argument for wildlife preservation, as advanced presently. Perhaps one day, but time is not on the side of wildlife. Better, I think, to conserve our energies and creativeness toward something of more immediate promise. The ordered universe of humankind cannot admit turns of conceptual conduct as anomalous as these.

RIGHTS AND DUTIES

This consideration flows logically from ethics and values. Lately a number of people have begun to investigate possible avenues toward the identification of "rights" in an inter-specific or even a systemic sense. If man has rights in respect to nature, surely the nonhuman has rights in respect to man. This speculation, perhaps too tenuous for serious contemplation by the confirmed liberal humanist, nevertheless has been usefully applied as a device with which to launch class-action legal process on behalf of a park, a forest, a water system, or something of that kind. Not yet a species, I suspect. However, I want to look at the question in the context of ethical behaviour by man toward wildlife. Does it contain the germ of an effective argument?

All concepts of rights are rooted in the value system of a society. The array of values to which a society adheres necessarily dictates a further array of duties and responsibilities that apply to both individuals and groups. The duties and responsibilities of the citizen are accompanied by appropriate rights. You have the right to be treated ethically and morally by your fellows, and at the grossest level you are also protected legally.

So we have the ethical structure of a society: you assume an obligation and receive an appropriate or concomitant right. If you work productively you eat well. Indeed you have the right to eat well nowadays, whether you work productively or not, which illustrates another contemporary value. Also children, the aged, and the insane have rights irrespective of any duties. The rights of "incompetents" spring from society's perceived obligation to them regardless of their productivity. That obligation springs in turn from a more fundamental value having to do with individual human lives.

We sometimes forget that every time a court or a legislature – or even custom – confers or confirms a right in someone, someone else's right is nibbled at: the right of women to equal employment opportunity is an infringement of the freedom of the misogynist employer; the right to make a profit is at someone else's cost; the right to run a motorcycle or a snow-mobile reduces someone else's right to peace and quiet in his own backyard; the rights of embryos impinge upon the rights of the women who carry them. And so on.

As far as wildlife is concerned, our right to consume it now restricts subsequent generations' rights to consume it then. This is the essential argument of self-interest, and of "wise use." And, as it turns out, of ethics. Nothing can be comprehended outside the realm of human interest, because that is where the human perceptual apparatus stops. The thought of any inherent right in the nonhuman, which by definition would mean the restriction of a human right, is impossible in our culture.

The notion of right is untenable outside human society. One simple example will do. Leopard eats gazelle eats grass eats soil nutrients. If the gazelle had rights with respect to the leopard, the leopard would starve. If the grass had rights with respect to the gazelle, there would be no gazelle. Everything would shudder to a grinding halt forthwith. It does not work that way. The individual blade of grass, in its final ecstasy of self-actualization, knows that its roots are alive and well. Rich, nutrient-packed leopard droppings will be forthcoming shortly, and tomorrow there will be a new blade of grass.

A hierarchy of rights in nature would be a one-way street. The leopard *seemingly* has a right to the gazelle which *seemingly* has a right to the grass which in turn has a right to the soil which in turn has the right to and will eventually consume all of them. But do these rights actually exist? I think not, at least not so long as right implies concomitant duty. The only "duty" any living being owes is to the recycling machinery itself, and that duty is absolute. No one is exempt.

But that is a human perception only; it is not a conscious perception in nature. God sees the little sparrow fall in certain and secure knowledge that it is going straight into the hopper for feeding back into His greater system. He owes the sparrow nothing; the sparrow owes Him nothing. The only affront to God would be the sparrow's unilateral assumption of immortality. and God knows that is not in the greater scheme — not for sparrows nor for anyone else.

For us to argue that wildlife has inherent rights that should be respected by man would be just as faulty as the long-established and prevailing assumption that man has total right of access to wildlife. The sheer raw power of man makes it appear as though nature were acquiescing to some human right, but that is an appearance only. The only way in which nonhuman rights could be taken seriously would be to identify the appropriate duty in each transaction: a duty on the part of wildlife to man, a duty on the part of man to wildlife. Neither is sensibly thinkable.

The "rights of nature" cannot be a useful argument for wildlife preservation, because like so many of our other

arguments, it turns upon itself. Nature knows no duty, therefore it can own no right. Still, as a stratagem against the time
when the hierarchical concept of our relationship with the
nonhuman is done with, it may have utility.

None of the traditional or even contemporary ethical
arguments for wildlife preservation has yet been shown to be
of the slightest practical value. In all ethics there must be the
fundamental assumption that the underlying values, beliefs,
duties, and obligations are fully, mutually understood,
accepted, and shared. In speaking of ethics in the nonhuman
context, we are jabbering into a void. Nature does not *need*
ethics; there is no one to hear.

SPIRITUALITY

Here we have the polar opposite to the icy rationality of the
argument of self-interest. Spirituality is less an argument,
more a persuasion. It is individual and deeply personal and
emotional. Conservationists have rarely used it in a broad
way. They have preferred to attempt to present an ultra-
logical case, with varying success. Usually the nearest we have
allowed ourselves to drift in this direction has been through
a kind of inspirational posture which, at times (especially at
the individual level), may become "spiritual" in a very loose
sense. There are frequent allusions to other powers and other
dimensions, usually couched in generalities such as "the
system," "the greater good," and so on. There seems to have
been some reluctance to attempt to exploit the latent spirituality that is in every human being.

There are many reasons for this and perhaps they do not
require elaboration here. Wildlife conservation groups and
organizations tend to be highly heterogeneous in their com-
position. A substantial number (probably most) naturalists
are not churchgoers (no, I do not confuse church-going with

spirituality), but virtually all of them have a deep emotional and spiritual feeling for wild nature in all its forms. I think conservation organizations have tended to shy away from any appearance of "using" this spirituality, even though between individuals and in small groups there is abundant sharing of the personal emotional experience. It would not be good policy on a public or mass basis, it is believed.

It certainly would not be good policy were the general public to continue to confuse sectarianism and other institutionalized believing for spirituality. In at least some quarters the wildlife preservationist is already seen as a kind of crank; to add to this any additional possibility of sectarian crankiness, as it is known in North American society, would be suicidal. Better, from a policy point of view, to leave the spiritual end of things, including "religious" experience, strictly alone. This is what we have done.

Even so, I cannot help wondering what the possibilities might be were we to uncover some germ of potential for wildlife preservation in the canons of one or another of the existing churches, of whatever stripe, for possible alliance. Not bloody likely. It is rather like politics. All conservationists have wailed and gnashed their teeth over the utter absence in the major political parties (or even the lesser ones) of any genuine commitment to conservation in the broadest (environmental) sense, much less in the more particular field of wildlife. Some of us have speculated whether we might actually form a political party dedicated to the advancement of the human good through environmental awareness and all of its appendages. (You would lose all your deposits in an election, but you would have had free time on television.) This, however, would only serve to alienate the rank and file of other political parties, and would be self-defeating. No doubt the same thing applies to organized religions; the conservationist has no wish to polarize spiritual positions.

There are few if any "all-purpose" appeals from the spiritual point of view, so we might well ask why not. The points I have raised are purely political or strategic, and they would not stand in the way if one were absolutely convinced of human spirituality as the route to wildlife preservation. Let's try to see, then, what it is about spirituality that has any relevance whatever to wildlife conservation. I will look at it under two categories.

NATURE IN THEOLOGY

This is not to be confused with "naturalism." There is little to be gained for wildlife by charging about the forest in beads and loincloths. Here I mean the role of nature and the non-human in the pervasive and important churches in our society. Apologists for each will no doubt produce reams of rationalizations to dispute me here, but I shall stand firm. I am particularly interested in the failure of organized religion to contribute one sausage's worth to the advancement of the cause of wildlife. On the surface of it, it does not make sense. Then again it does.

Many, including myself, have had things to say about the historic contribution of the Judaeo-Christian tradition to the estrangement of western man from his own biology, his alienation from his own being, and to the concreting of the notion of man's "dominion" over nature in the very foundations of western thought. This charge has been adequately borne out, and I do not intend to pursue it here. There are many sources in which you may trace the origins and evolution of the theological givens in our culture. Their effects have been catastrophic from a wildlife point of view. But all that is history, you could say, academic nitpicking. What about now? I shall address the now in the conviction that there has not been any perceptible shift in any branch of traditional western religion with respect to the relationship of man and nonhuman nature.

What interests me mostly here is the apparently wilful abdication of what I would perceive to be one of the prime roles of any church: to recognize the value to the human spirit of our life context. Nothing, it seems to me, could be more fundamental. If organized religions are to serve any purpose at all, surely it would be to purvey something of spiritual value here on Earth and now. What could be more appropriate for them than to attempt to understand the nature and spirit of *belonging*?

They are missing a heaven-sent opportunity, I think. Not only could they become "relevant" but also they might recapture their share of the market from some of the more zany persuasions that have drifted into the spiritual vacuum. Presumably it has not been possible for the churches to move with knowledge and philosophy through the last two hundred years because they have been unable to modify or temper or open their orthodoxies toward the acceptance of a universe that after all might not have been designed around man, nor indeed designed at all, a system that might not be hierarchic, a mode of interpretation that might not be dogmatic, nor even rational, a human being that might not be merely three-dimensional, a time that might not be linear, a cosmos that might not require causes, an order that might not be orderly.

You might well say that the influence of the established churches has been disintegrating for such a very long time that I am really setting up a non-issue here, stoning stained-glass windows, by suggesting the churches as serious factors in any sense. My feeling is that so long as cultural lip service continues to hold the churches in awe and respect, and so long as their world views continue to be the western world views, they are important. In spite of their apparent almost total inadequacy in "real life" issues, their past influence has been immense. (In some ways it still is; in redneck politics, for example.) Were the churches able to see and act upon the obvious as opposed to the theologically decreed needs of people, they might find a contributory role in society even yet, and might – just might – be of use to wildlife conservation.

Most naturalists tend to be pantheistic in mood if not in expression. There are no practising animists of my acquaintance, but there are few outright atheists either. There is, however, a deep and pervasive skepticism about institutions that pretend to hold answers or formulae for human problems while at the same time seeing man as outside biological process, and non-man as beyond the pale. On the other hand, there are a great many people like myself, who are waiting, with relatively open minds.

The shift in orthodoxy that would be required to make a church acceptable in the present day and in the present context would not be as radical as the bishops might fear. This is not a call for a crash program of arcane casuistry by the theologians, but rather a suggestion that there is an opportunity here to consider both the aesthetic and the ethical problems I have already outlined, on the basis of revelation well in hand. Man is a biologic being; a remarkable one, certainly, but still a biologic being, subject to the rules and regulations that apply to all beings in the biosphere, which is itself a biologic being. But there are opportunities as well as rules and regulations. These opportunities are lateral, however, not vertical; they are democratic, not theocratic; they are now, not in future; they are real, not mythic; they are individual, not institutional.

If it should turn out that the churches are in the business of *serving*, as opposed to protecting a vested interest in antiquarian antibiosis, they could yet be very positively important. If their call is indeed to serve man (or God), then their charge is to serve man and the environment from which he will always be inextricable. If, however, they persist in separating man from his living context, they can only exacerbate the contribution they have already made to the ecologic nightmare. (The technomachine is daily blessed by churchmen most high.) On the evidence, the likelihood of a change is not great. The likelihood of one coming soon enough to make any difference is even more remote.

Organized religion appears to have little or nothing to do with spirituality, which is after all an individual and not an institutional affair. It should be pointed out, however, that nature *does* have a role in theology. Nonhuman beings *do* have a role in the pursuit of inquiry into the natural history of God and other religious absolutes, if that is where your inclination takes you. I know of no formalized western theology, however, that even admits the kinds of questions that are raised by ecologic insight, much less begins to face them. The positive side of this is that ecologic insight gives rise to questions that require a spiritual investment for their open-minded pursuit that is qualitatively different from that dictated by any ecclesiastical dogma. It is much more difficult, but it is infinitely more rewarding, because it is about living. Living as experiencing is discussed in Chapter Four. Here let me merely say that I know of no even marginally useful argument for wildlife conservation that has been put forward by any of the major churches, and I can think of no honest argument that could be accommodated within their present positions.

Here we have a much more interesting issue, provided you do not read the word "mysticism" in any pejorative light. If you do, there is nothing to discuss. Mysticism has been around for as long as people have tried to label the kind of spiritual experience that is not satisfactorily explained either through approved orthodoxy or rational process. There are "religious" (here I mean truly spiritual) experiences that seem to arise out of individual intuition or contemplation, and that cannot be classified, labelled, and pigeon-holed. They occur in the quiet privacy of our minds. In others, we describe these experiences as "mystical," meaning "mysterious." We forget that there is no regulatory onus on the experiencer to explain his experience, nor even to discuss it. We also forget that such experience seems mysterious mostly because it cannot be force-fitted into convenient categories of common (shared) experience. Its whole reality is in one individual only. This the naturalist knows – profoundly.

There are all manner of posturers and charlatans who for various reasons like to advertise and even to preach their mystical experience. They may do this while assuming an aura of serenity that is meant to convey the message that they are privy to knowledge that is not ours. It should be obvious that the naturalist is anything but serene. Neither is he privy to the slightest scrap of conventional knowledge that is not available to all. His is nothing of the occult or arcane, nothing of the mystical.

There is however a very special, very personal, and at some times highly spiritual quality about the *experience of the naturalist*. He does not tend to advertise or proselytize the experience because of its intensely individual nature. But, paradoxically enough, he has an urge to transmit *something* to others. In its most developed form the urge is not to communicate fact but merely to share feeling. In a political sense it would be to the greater good were more people to partake of the experience, but in the glow of enthusiasm this, I feel sure, is not the motive. The joy is too great, too overwhelming, to contain, so it bubbles up. If you have not experienced it, you will have to take my word for it. If the naturalist does not know something you don't, he most certainly *experiences* something you don't. He is not serene. He is excited.

This experience defies communication and defies institutionalization and dogmatization. Since it cannot be rationalized and quantified, one cannot package it. Most especially, one can't politicize it. For this reason, I was one conservationist who was not especially happy when the "environmental movement," if I may call it that, was unilaterally adopted by the "counterculture" of the latter 1960s. This did wildlife preservation no good. Conservation in general tended to become identified with the attempts of the counterculture adherents to polarize opinion politically and otherwise on a wide range of issues. I believe that conservation, identified as it was with "environmentalism," was the loser in that instance, and it has taken it longer to become "respectable" again. To this day in the seats of power, which means in the corporate boardrooms and the highest echelons of government, "conservation" has too often meant "environmentalism" which has too often been interpreted as hippies and radicals intent on the indiscriminate overthrow of all things. This, whether we admit it to ourselves or not, it still a serious perceptual and thus operational problem.

This is not to downgrade the contribution of that period in history to general awareness of ecologic and related problems. However, as much of the credit does not belong with the "counterculture" as some have suggested. It had been coming for a very long time – since the 1930s really – and in the self-interest sphere, to be sure. What the counterculture did do was bring concerns of this kind to the attention of the young, who theretofore had been mostly absent from conservation councils. They still are. My regret is that the environmental concern was so often expressed in doctrinaire terms, as part of a larger ideological "package." It did wildlife little good.

Since much of the flavour surrounding the counterculture was selected from eastern cults, philosophies, and religions (somewhat indiscriminately, it seems to me), in more traditional eyes there was much of the "mystical" – again meaning "mysterious" – about it. (And in this case, "mystical" *did* become ever so slightly pejorative.) That there was spirituality there, at the individual level and at least sometimes, I grant. I do not however perceive that there was much in it of what I feel to be the spiritual experience of the naturalist. It was contemplative, to be sure, but it was inward. The universe of the counterculture was as homocentric as any designed in Athens or Rome. As it extended beyond the individual, it extended to other (human) individuals on a parallel "trip." Like most if not all romantic movements, it had little to do with living nonhuman nature. Its "environmental" aspect was social and cultural, not ecologic.

Lately, a useful distinction has been drawn between what are described as the "deep" and the "shallow" environmental movements. I would characterize the counterculture and much of contemporary "environmentalism" as the latter. The former is applied to the more reflective philosophic examinations of ecology and man, which although not the exclusive province of naturalists, tend to arise out of the naturalist's experience. Because the spiritual experience of the naturalist is so vividly personal and so resistant to classification and communication, there is small chance for us to "reduce" it to a form suitable for communication. This in spite of the long tradition of "nature writing" I mentioned earlier. There is much in our literature that persuades the reader very close to the qualitative experience of nature. Oddly enough, the best of it has not been written by "practising" naturalists, or at least not by people primarily known as naturalists. A Yeats, a Lawrence, a Coleridge, an Eliot, a Frost, or a Whitman – yes, a Poe – brings me ineffably closer to the core of *experience* – and thus of the nature of things – than most of the purple, adjectival, mawkish outpourings that I find on the "nature" shelves. That is because the best of them are talking about *nature*, not projected-reflected narcissism. About living in all of its dimensions. They are sometimes described as "mystical."

Mysticism is a private and individual experience. So too is the spiritual state of being that is wildlife preservation. It is not communicable in the usual ways because it cannot be presented through logical explication. It is non-argument, non-rationality, non-sense.

It seems to me that spirituality, whether in the form of traditional religions (western or eastern is of little consequence) or in that of individual "mystical" or intuitive experience, has not yet developed as a workable, reliable, or effective argument on behalf of wildlife preservation. That spirituality of some indefinable sort is close to the root of the naturalist's ethos is indisputable. We have not yet found a way to market it. We are not even certain whether we should try to market it.

THE CENTRAL ASSUMPTIONS

In all of our arguments for wildlife conservation there are certain basic assumptions. Some are stated; more are implicit. Since I am concerned with the *effectiveness* of our various arguments, at the risk of seeming repetitive, I want to do a brief review. It will begin to isolate and shed light on our assumptions. In many cases it is the assumptions themselves that ultimately determine the persuasiveness of our arguments.

Self-interest (prudence) assumes, essentially, that the human being is capable of acting in anticipation of things to come, of profiting by experience, of planning on the basis of knowledge of possible outcomes. This is the assumption that people will not rebuild their houses in the flood plain after the deluge, or that they will scrupulously save and invest now against unspecified future reward at an undetermined date. In brief, it is given that we are willing and able to act in our own future interest, even though the form of that future is by no means established.

It assumes further that wildlife conservation can be shown to be in the human interest. This argument requires the assumption of wildlife as a commodity that can be perpetuated in the service of man. Wildlife as a renewable asset assumes man as proprietor and decision-maker. An additional underlying assumption is that in each case a quantifiable payoff in cost-benefit terms can be demonstrated.

The human right of total access to wildlife for entirely human purposes is assumed throughout. Wildlife remains a human commodity whether man is principal or agent, emperor or steward. The utilitarian assumption runs through the entire roster of "wise use" arguments, and, albeit less visibly, in the "quality of life" appeal.

In the latter, the payoff is qualitative rather than quantitative, but there remains the assumption of wildlife as an asset in that man is the beneficiary whether or not he consumes wildlife on the spot. The investment metaphor holds, with the implicit assumption of right of proprietorship, even in the realm of aesthetics. Again, a future orientation assumes our willingness to pay now, benefit later.

The "ecocatastrophe" argument, the ultimate in the flood plain set of assumptions, assumes that you can actually scare people into acting in their self-interest, on, when it comes to wildlife, very flimsy evidence. The implication that wildlife can be shown to be in the human interest is rarely proven out.

Ethics as applied to wildlife assumes that a code of human intracultural behaviour in the social interest can be extended interspecifically. This assumes in turn not only interspecific reciprocity (which implies some mode of communication) but also that there is a wider social unit than man's, the members of which are able to act in the best interests of that unit. It assumes the human capacity to transcend the human interest and the nonhuman capacity to recognize not only its own but also the human interest.

In a very real sense the ethical argument requires that we see the nonhuman world in terms of the liberal egalitarian ideal: that the individual being is sacrosanct, that rights and concomitant duties, equitably proportioned, will serve the long-term higher purpose, and that "equity" can be applied to natural process. In brief, it anthropomorphizes the non-human world in order to include it in a human ethical code. All of this assumes an interspecies *summum bonum*; it assumes further that man, as proprietor, has both the right and the duty to *extend* an ethical code to those of lesser hier-archical standing.

It seems that arguments for wildlife conservation, whether couched in terms of human self-interest or in terms of human ethics, must either assume or result in a hard line of distinc-tion between man and non-man. There appears to be the necessary setting up of a duality so that a "client" relationship may be perceived. There is the further assumption that people are able to see and act on behalf of future generations of people.

Spiritual arguments would like to remove the man/non-man distinctions, but those couched in theological terms can do nothing about the assumption of a man-centred universe. Although wildlife may be seen by the churchmen in an entirely benign way, man must be assumed as guardian or patron at the very least. God's steward on Earth enters into a landlord-tenant relationship with non-man. The assumed hierarchy is essential.

Mysticism, as we have seen, is an individual matter. Since there is no agreed-upon mystical argument for wildlife preservation, there are no prevailing assumptions.

Argument itself must assume communicability. Also it assumes logic, reasonableness, and persuasive evidence. It assumes dialogue.

Obviously, the difficulties raised by our assumptions are formidable. Whether tacit or acknowledged, they colour our entire process of reflection, and they cannot help but bring a fundamental bias to both our deliberations and our stated case for wildlife. Virtually all of our assumptions tend to reinforce and legitimize those concepts, so deep-rooted in our traditions, that have not only brought about the present condition of the life tapestry but also stand so implacably in the way of wildlife recovery. I should like to address some of those root problems directly.

THREE

The ROOT PROBLEMS

There is no assumption – just as there is no individual or species – that arises spontaneously out of nowhere. Like any other idea, an assumption, whether articulated or not, has an origin and an evolution and is able to persist at any given period only because the prevailing environmental climate permits it. Were it not "adapted" to the time and place, it would not be there.

In this chapter I look at some of the roots of those assumptions that seem to cloud our vision and inhibit our action in wildlife preservation. There are many of them, but they seem to coalesce into three general categories. There are those mechanical, operational, in-the-now problems that I shall characterize as "functional"; there are those that, arising from our social and cultural tradition, are ideal or "conceptual"; and there are those that spring from both our biology and our accumulated experience, and are in general "perceptual." All inter-relate and interact. Conceptual notions clearly help to shape both functional and perceptual difficulties, just as our perceptual capacities allow us to create and view both concepts and operations in certain ways.

THE FUNCTIONAL PROBLEMS

The keystone of western civilization is the assumption of absolute truth. Our theologies, philosophies, science, and law, among other endeavours, require for their existence if not the tangible presence then the certain promise of unequivocally true answers. We behave "logically," under the over-riding assumption that just as there are reasons for all things, so all things – ultimate facts – are available to pure reason.

I shall return to this issue (which is of course a conceptual one) later, but I raise it here because it is at the heart of one of our most recalcitrant functional problems, and that is the matter of reasonable evidence. As we have seen, it bears heavily on the "ecocatastrophe" argument, and directly or indirectly on every aspect of the entire spectrum of "self-interest" appeals. In practice, of course, the usual absence of incontrovertible evidence is not a problem; the doctrine of "reasonable doubt" allows us plenty of freedom to leap to the conclusion that we favour. When it comes to the self-interest, especially if the argument is delivered in the form of a threat, nearly all of us want to be shown, beyond a shadow of a doubt. Ask any smoker.

The futurologists know better than the rest of us how extraordinarily difficult it is to convince the public or any significant part of it in the matter of futures that may be something less than appealing. As a result, most of them tend to bend, warp, or at least soften their predictions, or at the very least to leave available trap doors for the sweeping out of mind of any threatening implications.

Proving it in the case of wildlife conservation is not easy — in fact it is virtually impossible. (You should not be misled here, into concluding that wildlife plays any part whatever in technocratic planning and future-casting; I speak here of the futures outlined by ourselves, the conservationists.) Clearly we cannot *prove* anything in future, because the future has not yet happened. The best we have is hindsight reference to what has happened and is now happening around us. Unless there is a greater emotional investment in the future of wildlife than I can detect around me at the moment, mere uncertainty is a crashingly weak and insufficient argument. It leaves a gap of "reasonable doubt" big enough to drive the technomachine through. Even were there proof in hand, of course, we would have a perceptual problem, but that is another matter.

The unavailability of proof and thus of ultimate truth is a severe limitation to any conservation argument, but nowhere is it quite so bothersome as it is in the sorry burlesque that is "environmental impact assessment" from the ecologic point of view. I have already singled out this travesty for special mention; here we must attend to one special aspect of it. Now we are concerned with prediction. Many branches of science are quite comfortable with prediction (if they could not handle measurement, including extrapolation, they would not be sciences), but ecology is not in that position.

Although ecology has long since shed its swaddling clothes, it is still young and has much to learn. The more it learns the more it discovers that it does not know. In fact, "pure" (as opposed to "tame") ecologists claim to know only one thing for certain: that they will never know one thing for certain. I can think of no "hard" science so helplessly adrift on an endless sea of variables, their process relationships so complex that no individual mind can either encompass them or corral them for computer counting. Ecology does not know what its variables *are*, much less how to project them. When you remember that ecology is site- and situation-specific, forecasting becomes so vexatious that "ecologic prediction," though eminently useful jargon for the technocrats, is no more than a nonsensical contradiction in terms. Ecology cannot predict in the scientific sense; it can merely describe.

Historically, however, and for reasons which need not be elaborated in detail, wildlife conservation unfortunately found it necessary to accept the burden of proving the effect on wildlife and wildlands of various kinds of human activities. If the prediction could be quantitative, so much the better; all is possible in the rationalist-scientific world. Along the line it had become abundantly clear that there was little more than the most meagre constituency for "wildlife for its own sake" in the corridors of power, and rather than attempt to build that constituency whilst holding the philosophic line, conservation (at least in part for purposes of communication) chose to enter the mod arena of cost-benefit. That made us respectable. Now we could converse with civilized folk.

Thus the series of "self-interest" arguments came into being. If you won't save something for the sake of saving it, then here are some reasons. Unfortunately the juggernaut that is the current socio-economic order is fuelled on hard data. Wildlife ecology just does not have them. All that an opponent in an impact hearing has to do is ask for conclusive evidence. More data. Like good little soldiers, we jump right back on the treadmill.

Having painted itself into a corner by attempting to quantify the human interest in species of wildlife, conservation desperately adopted the "ecocatastrophe" approach. Its effectiveness is well-known. When it accepts the burden of *proving* – of giving *evidence* – conservation effectively defeats itself. Having in mind the limits to ecologic prediction, wildlife conservation can never present an airtight case. And of course industry, commerce, and the growth imperative are required to *prove* nothing. Their goals are part of the belief system. Wildlife conservation is not.

There is another grave functional problem, also having to do with the future. Both the "stewardship" and "ethics" themes made specific reference to our duty to future generations. Other arguments contain the same injunction at least implicitly. There is an assumption: not only do we owe an obligation to the future, but also we are able to recognize and act upon our obligation to the future.

The fact seems to be that we have very great difficulty in entertaining any tangible feeling of responsibility for our descendants. I know that I can "identify," as they say, with my children, at least in the here and now, and for a stone's throw into the future. I am not at all confident that I can even *feel*, much less see, my way through their lifetimes. If I ever have grandchildren, I may be able to have a similar identification with them, but I will not be certain of that until I have seen and touched them. Certainly I cannot empathize with them now. (Indeed I feel much closer to the Peruvian cock-of-the-rock, a bird I have never seen in the wild, but that I know *exists*.) I am not at all confident, in other words, about my ability to feel any emotion at all in respect to future human generations, much less any responsibility. I think that this incapacity is not unique to me. I believe this is a true and real characteristic of all of us, no matter how often or how portentously we rhetorically indulge those not yet born.

I know very well that we *say* we can assume this obligation, but *can* we do it? I seriously doubt that as sentient beings we are really capable of abstracting that *idea* called "future" into a sufficiently palpable reality to allow us to accept responsibility of any serious kind. This is not meant as a criticism of our finer attributes; although I cannot prove it, it is a comment on something I suspect about our biology. Why, after all, *should* we be able to do it? "Future" will forever remain a concept, not a tangible; it cannot be an external stimulus, not ever.

Robert Heilbroner says it is "the absence of just such a bond with the future that casts doubt on the ability of nation-states or socio-economic orders to take now the measures needed to mitigate the problems of the future." I believe we have a major functional difficulty with this, not only as individuals but also as communities and societies and a culture. Although all sensitive folk castigated the senior publishing executive who asked, "What has posterity ever done for me?" I suspect he was more accurately interpreting the public pulse — the sociocultural heartbeat — than we do-gooders have been doing.

This brings us ultimately to that which is very probably the gravest of our functional problems, and that is the present (not the future) perceived self-interest of our society. I know of no argument for wildlife conservation that can so much as find itself on the agenda if the latter is cast in terms of contemporary human priorities. The poor are prevalent. I am not going to discuss the underprivileged overpopulated tropics and sub-tropics, because there seems simply no means of expecting a rational wildlife conservation argument there. (In such countries, by the way, the elites have no more interest in wildlife preservation than do the poor; it is not entirely a matter of economics.) In our own society there are also the poor; even more important is the industrial growth economy, which is not only a social phenomenon but also an article of faith. As long as profits must increase (who receives them is of no consequence here), and as long as yearly growth in the production and consumption of fabricated commodities is the sole measure of the health of a society, the blockade to wildlife conservation will be insuperable. There is little doubt as to which values come first, once things come to the "hard decision."

Functionally, then — which is to say in everyday practice — wildlife conservation arguments are expected to be expressed "rationally," quite in spite of (a) the limits to ecologic prediction; (b) the present locus of the burden of proof in such issues; (c) the apparent limits to effective human foresight and emotional commitment to the future; (d) the perceived socio-economic priorities of our society. I have not yet become quite so paranoid as to see this as a "conspiracy," but it has many of the earmarks. Let us merely say that it is one of those "Catch-22" situations that seem to emerge so predictably in virtually any aspect of the man-non-man relationship.

The functional problem, of course, turns out to be the belief system.

THE CONCEPTUAL PROBLEMS

These have to do almost entirely with the western belief system. We live in a bottomless alphabet soup of givens and assumptions, and although each of us has his or her special little group of variances from the norm, relative to the whole they are small indeed. Outside the institutions for the insane, there are very few fundamental concepts that are not shared across society. Strange and wonderful though many of these may be, we can only touch on a handful. For example, there is no belief more basic than that concerning man's absolute power and authority over the nonhuman. As we have seen, from it have sprung most of the assumptions underlying our arguments for wildlife conservation. A corollary of this is the belief that in the order of cosmic and Earth imperatives the continuity (immortality) of the human species has the ultimate prior- ity *over all other things*.

Although it looks as preposterous as it properly should when you set it in type, the belief is very real. This conceptual precedence over nonhuman animate and inanimate "nature" is a cultural mutation, not a biologic one, and it has not neces- sarily been common to all human cultures. It so pervades recent and contemporary thought, however, and has become so concreted through generations of tradition, that it is almost as though it were genetic. It is as fundamental to the thinking of the "developed" world as the notion of linear time.

Quite simply put, the conceptual power structure consists of an imagined evolutionary hierarchy in the Aristotelian mode, with man as the apex of a pyramid, as the topmost growth bud of a tree, as (for modest men) the penultimate link in a grand chain leading from the dust to God. At pro- gressive levels downward from our platform there are ranked and arranged those beings of "lower development." Thus a baboon is more "highly" evolved than an albatross, which is more "advanced" than a butterfly, which outranks any giant redwood. And so on, down the line.

The obsessive practice of ranking, grading, and assigning relative "worth" or "importance" to things (especially non- human things) is ancient. It is undertaken mainly, I think, in the interests of self-assurance. It is a most effective means of sustaining the conceptual power structure over the non- human. In another context it is sometimes known as the "kick-the-cat syndrome."

Evolutionary ranking through life continuity is justified on the assumption that evolution, up to the present time, has been purposeful, deterministic, and (most especially) progressive. And it is now *over*. Man is the final and only absolute parame- ter of evolution. All things were set in motion toward a single goal – the shaping of the ultimate creation, the Reflective One. Were the whole deliberate process to begin again, it would necessarily culminate in the same form. Evolution could have no other "purpose"; no other would make sense.

THE CONCEPT OF "MAN/NATURE"

The conceptual human power structure over nonhuman beings is supported and sustained by an even more radical belief, that of the fundamental difference between man and the nonhuman. Not just between man and chimp, or man and australopithecine, or man and bird, but the basic qualitative gulf between man and *all other sentient beings*. Man is in one compartment; all other life is in another. Our unshakable insistence on this has familiar manifestations. It is as though man were in some mysterious way abiotic, as though in some profound sense man were outside biologic and ecologic process. Thus we have "man *and* nature," "man *in* nature," "man *of* nature," "man *from* nature." The conjunction does not matter; the proposition is the same. The conceptual dualism is clear. Occasionally, in fits of desperate goodwill, we may substitute a slash "man/nature" in a well-meant attempt to indicate a bridging "interface" or ecotone, whilst all the time and often quite unwittingly underscoring the image of two discrete life systems.

Even when we do get around to acknowledging man as a biologic being, we appear to hold that he is quite a different *kind* of biologic being. Teilhard de Chardin, for example, seriously proposed man as a separate phylum from all the other chordates (animals with spinal cords). I am quite certain, however, that he saw man as outside the animal kingdom entirely. Thus we would have three life categories: plants, animals, man. Although we do not very often get around to pronouncing things like this out loud, in general, this is the way western culture sees it. In spite of our precious rational process and in spite of our cherished scientific objectivity, we continue to maintain an absolute and unchallengeable distinction between man and the nonhuman. It has occurred that the firmness of this insistence may be one measure of the need we may perceive for justification of our overwhelmingly antibiotic actions.

Life is filled with ironies, and I detect one here. There *may* be evidence to suggest that we *are* different not in degree but in kind. It arises from what would seem to be the uniquely human capacity to conceive of interspecies dominance hierarchy.

THE CONCEPT OF INTERSPECIES DOMINANCE

Many kinds of animals live in social groups; many do not. Leopards are "loners"; lions live in prides. Wolves have a very intricate and interesting social system, and so do some of our closer relatives such as chimpanzees and gorillas. In such groups, we see individuals occupying relatively stable social positions in relation to the others, in a kind of "peck order" such as is well-known in every henyard. This system of organization is called "dominance hierarchy." Those who have carefully observed the workings of such organizations in wild animals have emphasized the positive value of the arrangement. It is adaptive, survival-oriented. In other words, it works.

Hierarchy seems to work for the group because it is a manifestation of what is probably one of the most fundamental characteristics of life — the drive to co-operate (the "compulsion to comply"). I am trying not to be "mystical" in suggesting that after a lifetime spent in the appreciation of nature, the single factor that appears to me to be more important than any other is *compliance*. I can very comfortably interpret ecologic interdependence as co-operation. When a predator takes one of "my" grouse in the bush, it is not a crime against me or against God. It is co-operation on the part of the grouse. There is such co-operation, effectively, at every level of the ecologic structure. This is the point made earlier in the discussion of "rights."

Those species that live in groups achieve co-operation at least in part by a hierarchical structure in which the status of the individual has to do not only with other individuals but also with perquisites and responsibilities. Social status ("rank-order") in those species is essential not only for the survival of the individual but also of the group. For one thing, it does away with the inevitable quarrels and conflicts that would arise from a more "democratic" arrangement. Quarrelling and fighting waste energy, and also take your attention away from matters of security. Most important, though, a kind of ritualized or conventionalized social behaviour eliminates any need for jousting, which if carried to extremes has rock-bottom survival value. Food, shelter, and sex are all secondary imperatives; they are the rewards of social compliance.

Rank-order in nonhuman social groups is always in slow flux as those at the top age and drop away and young adults mature. Also – and this is important – groups tend to stay at about the same size. We cannot get into a discussion of the mechanisms that appear to regulate populations, but it seems that there are in each species upper limits to the size of group in which proper social organization can be maintained. This upper limit varies from place to place and from species to species, but organization is absolutely essential. Zoos have learned by sorry experience not to put too many baboons together, for example, in a confined space. The blood-letting can be severe – much as in human cities. Such behaviour does not have survival value.

The *pattern* of social organization is peculiar to the individual group. It has been worked out, as it were, by that group only, and every individual in that group knows his or her place in relation to every other individual. It does not extend to neighbouring groups, which have their own particular organization. Groups tend to keep away from each other, and when wandering individuals happen upon a strange group, there can be turmoil until the newcomer is accommodated (or rejected). Not only are social patterns specific to individual groups but also they are not extended to neighbouring groups of *other species*. That aberration is reserved for man: here, human uniqueness moves into sharp focus.

We have a remarkable tendency to perceive (believing is seeing) cross-species dominance hierarchy in other animals. Naturally it would be very satisfactory could we find a precedent in nonhuman nature. So, we observe interactions between individuals of different species and leap to the conclusion that some species are "dominant" over others. We are forever seeing "aggression" in nature; we like to read malevolence into the behaviour of a meat-eating predator, or belligerence into the behaviour of a grosbeak at a feeding tray. We forget that the predator and the seed-eater are merely feeding, and that feeding is basic and urgent.

Granted, we sometimes see birds of different species competing, as it were, for possession of a prized nesting-hole. This we swiftly transpose into interspecies aggression (striving for dominance), when what it really is, is competition for a nesting-hole. It would take the wildest foray of the human imagination to interpret it as illustrating an inter-species social dominance issue, complete with mutually understood rank. But we do it all the time.

Interspecies acceptance of social dominance hierarchy would be impossible, at least in part because the multitudinous species involved would not be able to understand each other's conventional symbols of rank: rituals, signs, language. A peacock displaying in front of a goose is going exactly nowhere. In the simplest terms, communication is generally understood as the transmission and reception of *mutually understood* signals. You have a sender and a receiver and a message understandable to both; otherwise you do not have communication. The dominance hierarchy in a group of social beings is based on a system of signs and signals which, whether the individuals learn them or are born with the knowledge or both, are usually fully understood and acted upon by all concerned.

On the other hand, the conceived human dominance hierarchy is completely one-sided. This is not communicated, because we are the only ones who understand the message. There is massive transmission, but no reception. Our physical power is so great, however, that, at least to our own satisfaction, we are able to display our dominance whether or not nonhuman beings "get the message." Thus we are able to fool ourselves as to our role. One of our prime self-delusions is that nonhuman nature *accepts* our dominion, as part of the natural order of things.

Before we go further with this, a word on how dominance hierarchy seems to work in nonhuman societies. We usually describe the top dog as "Alpha." There may be individual rank-order down from there, or there may not, depending upon the species and the situation. In any case, except for the inevitable turnover through time, one thing we consistently see in patterns of hierarchy is their stability. This homeostasis seems to be maintained, at least partly, by the way in which both dominant and submissive individuals behave. Their behaviour tends to reinforce their positions in the scheme of things. Sometimes you can recognize Alpha in a henyard or a baboon troop by his or her appearance, or posture, but sometimes not. Very often it is much easier to *infer* which individual is Alpha by watching how the *others* behave. The way the subordinates act can often direct you to an individual who does not appear to be doing much of anything at all. The others are behaving *in relation* to that individual, who turns out to be Alpha. Watch an individual back away from water or food, or step to one side on a trail. Then watch the other one. You have found Alpha.

It's rather like trying to spot the human Alpha in a large crowd. You do not usually identify the dominant individual by his own carriage, but rather by the running about and breathless bowing and scraping and general obsequious behaviour of the people swirling around him. From this, you *infer* his presence, and indeed his qualities. Think how often it is that we infer a social hierarchical position in someone, not by his show of power, but rather by the show of subservience in his followers. We see many more displays of deference and submissiveness than we ever do of naked strength. Alpha does not need to do very much when power is constantly reinforced by his subordinates. He may indeed have greatness thrust upon him. Such is the case with a king, and it is also the case with a male gorilla, whose sheer size makes it quite unnecessary for him to do anything at all. Also, lest we forget, there is no shred of evidence to show that either the king or the Alpha gorilla ever *wanted* to be where they are.

In the course of studying individual interactions between members of groups of social animals, various ethologists have called attention to what they have called "inhibitions" against fighting. At the approach of the biggest or "dominant" male wolf, for example, another individual may roll over on his back like a puppy, or slowly move his head to one side, making his throat openly vulnerable to the slashing fangs of the large male. But nothing happens. Many like to conclude that the big animal is "inhibited" against attacking; the action of the subservient individual prevents the attack. They say that he is "inhibited" because they apparently presuppose that the wolf is by nature a bloodthirsty, slavering killer, and has to be prevented from indulging his true "instincts."

This interpretation, like so many of our perceptions of nature, is the victim of our presuppositions about nature. No doubt it would be a wrench to have to abandon "nature red in tooth and claw," but that is where an unbiased appraisal of what we observe must lead us. I know of no evidence whatever to show that the wolf or any other animal, including man, is innately savage. The one characteristic that seems to cross all species borders is compliance. The big wolf does not need to be prevented from attacking; he has no intention of attacking. Both he and the subservient wolf are complying with the social imperative against internecine fighting. Indeed, by going through this little ritual they are solidifying and reinforcing their particular social relationship. Nothing could be more destructive of the group integrity than to have one or more of its members injured or incapacitated when the time comes for feeding.

There can and will be fighting, however, when the appropriate signals of compliance are not forthcoming. In such cases Alpha appears to be triggered to action by some perception of disharmony (disorganization) which he swiftly moves to correct, in the group interest. It is true that there are occasional fights between apparently equally matched "dominant" males in nonhuman societies, but these are usually very short-lived, and, it seems, most often lead to early peace.

I am strongly persuaded to the view that in "real life," which is to say nonhuman nature, there is no such thing as dominance. Dominance may well be an inference to which we attribute reality, or, if you will, a human projection. It seems to me that what we anthropomorphically call "sub-servience" or "submissiveness" or even "obsequiousness," but what is in fact *compliance*, is the inherent drive that makes the biosphere go round.

If indeed "dominance" is manifested and hierarchy maintained at least as much by the actions of compliant individuals as by the actions of Alpha – perhaps more so – then the failure of nonhuman nature to receive our message of superiority, to accept our dominion, and to act accord-ingly creates a genuine problem. Here perhaps is a root cause of much of our cognitive dissonance with respect to wildlife. But since the one thing the Rational Being cannot do is leave the question open, we must *pretend* that nature is indeed complying, and is playing its proper reciprocal sub-servient role. And that is exactly what we do.

But it is one thing to pretend and another to make the pre-tence *rational*. We have a neat way of doing it. Since wildlife cannot or will not see us in a dominant light (we are not of their species, therefore we are socially irrelevant), all we have to do is see wildlife in a human light. Then everything falls into place. We do this by seeing wildlife *comparatively*. We judge wildlife species by human standards *in order to find them wanting in human qualities* so that they may be appropriately ranked and filed. Because our standards are specific to us, no other species can possibly meet them. Man is thus the rational measure of all things; the proof is universal, and the perceived hierarchy is firm. There is not a word of protest; we infer compliance. (Consider if you will the possible outcome were united gerbils of the world to declare gerbil qualities as stan-dards for determining interspecies hierarchy. Perhaps indeed they have done so, and live in total comfort and assurance.)

It is essential for the interspecies dominant to cast other species in his own image, not to see himself in theirs. The beloved dog is "almost human." To anthropomorphize the non-human is much more salutary from an ego point of view than it would be to zoomorphize ourselves. After all, we are but one species and nature is millions; it is more satisfactory to assign roles to those bewildering multitudes in the mono-specific human hierarchy. To attempt to do otherwise would be meaningless, because the only universal meaning is in human supremacy.

Thus it is that anthropomorphism becomes more than merely an irritating aspect of the Sunday supplements, the performing captive dolphin shows, and the "true-to-life" nature films. It is a fundamental need for the self-styled Alpha. Whether or not the concept is carried into action by way of overt force is incidental so long as "lesser" or "subhuman" beings are seen as appropriately subservient and submissive.

One of the more rewarding aspects of anthropomorphiz-ing is that it allows us to attribute to other forms of life certain qualities and motives which, so far as one can tell, are uniquely human, and then to act accordingly. There appear to be "good" and "bad" human individuals; there must then be "good" and "bad" species. Once a human attribute has been assigned to a nonhuman being, it becomes possible to per-ceive that being as an errant human, and to take appropriate remedial measures. We can bring to bear all human ethics and law against the fox who "steals" a chicken. We can rename the orca "killer" whale and carry on in suitable fashion. To see the great homed owl as a "cannibal" and a "bloodthirsty murderer," as some of the nicest people have done, is to justify appropriate punishment. Conversely, the sweetness and light of the tranquil garden serve to reflect the poet's self-satisfaction. This is part of the same process. To cast the rest of nature in our image allows us to use human criteria in the valuation of the nonhuman, absolutely guaranteeing the perpetuation of the perceived scale of cosmic priorities. No shell game was ever more remorselessly rigged.

Anthropomorphism has a long, long history. On the record, as we have it, it turns up first in some of the scrawlings on cave walls that show creatures part human, part nonhuman. Later we have supernatural beings, and inevitably these begin to assume human form. Much later we have Mediterranean myths having to do with people in animal form, and so on. One anthropomorphic God turns up a bit later. Then it is on to stereotyping, various species being vested with sundry (human) attributes, which in turn allow the valuation of relative worth.

One of the very best examples of the uncanny way in which we can twist things around occurs in the present day. In describing the Alpha male of some primate societies, some anthropologists have gone so far as to call him the "despot" of that group, and to refer to social rank-order as "social despotism." The extraordinary thing is that this is nowhere visible; it is entirely inferred. Believing is seeing. The conceptual-perceptual overlays are interesting.

In practice, the function of the human despot has a special twist. In other group hierarchies, Alpha provides defence and reassurance, permission and protection. He carries out many important responsibilities; they sometimes cost him his life. The self-perceived human Alpha, on the other hand, persists in denying those in "lower" positions the largesse of his tenure. At the very least, he is imperially indifferent to the well-being of the infrastructure. He is the sole beneficiary of the organization. Again, the flow is one way. This, over the long term, does not have "survival value."

The survival value to the group (and thus, ultimately, to the species) of dominance hierarchies in nonhuman social beings is basic. A nonhuman Alpha who did not carry out his several obligations to the group would soon be selected out. Those dependent on him would no doubt be selected out also. This is to say that the survival value of a dominance structure that is incomplete (one way, non-reciprocal) must be severely limited. We like to forget this.

I am quite confident in surmising that one of the reasons that nonhuman beings do not accept our unilateral dominance position (whether or not they might otherwise interpret the message) is that they probably cannot *conceive* of interspecies rank-order. They cannot conceive of it because they don't need to be able to. They — especially those that are highly social — play out their individual lives within the internal structure of their individual groups. They do not appear to require the constant, repetitive reassurances that they might otherwise derive from projecting their own hierarchic structure onto neighbouring groups of their own species, or of other species. They are self-contained, mature, "adjusted." Nonhuman beings, so far as I can tell, are totally devoid of the gnawing ache of "group ambition," much less "species ambition." They appear to me to be whole.

Man appears to have a piece missing. Perhaps that piece was the price, paid long ago, for the uniquely human quality of ambitiousness. For in our species, the ambition of the individual becomes the ambition of the group, the tribe, the species as a whole. Interspecies dominance becomes global — becomes cosmic dominance. The drive toward Alphaness becomes a species imperative. Dominance becomes the reason for being.

THE CONCEPT OF HUMAN IMMORTALITY

I think it is permissible for us to speculate that nonhuman beings cannot conceive of either interspecies dominance (ambition) or of interspecies submission (fatalism) because they are not hagridden by our peculiar brand of fear. Man is the fearing animal, the anxious animal. Species ambition in man seems to be the inevitable consequence of fear, or, at the very least, of chronic insecurity. Anyone who requires reassurance in the continuing massive doses western man demands, would be said to have a problem. I believe we have a problem.

It is generally accepted that the most basic human fear is the fear of death. This may or may not be true of the individual, depending on his age, but it appears to be profoundly true of the species as a whole. If there is one single human imperative it is this: *man must not be subject to extinction*.

The very notion of extinction as applied to man is beyond decent consideration. I have found that the most shocking and offensive idea I can offer to my humanistic colleagues is that when man *does* become extinct, *he will never have existed*. There will be no one, no thing, on Earth to recognize our rubble for what it is, or to make any connection between it and anything that may have passed, so it will never be known that we were here. This, I am told, is the ultimate "pessimism"!

If I am right about the human immortality imperative, then human fear is the fear of natural selection: as individual, group, population, race, species. It is passing strange. In full knowledge of the splendid procession of diverse and beautiful forms which preceded us and now accompany us through Earth time, why should we fear to be part of it, I wonder? What is it that compels us to be first, I ask? What is it that impels us to be *last*?

It seems clear that our perceived dominance would be sorely affronted by an eventual turnover in Alpha incumbency. (That there must be an interspecies Alpha is given, it appears.) We believe that we are sufficiently in control of and qualitatively different from the natural infrastructure, that we quite clearly resent the possible application of life rules to ourselves. Since the goal and the purpose of evolution were to produce man, it would make no sense for the ultimate creation to thereupon become extinct. That would contradict the rationale for the entire life process. World without man is quite literally unthinkable.

Also, it would seem to be something less than fitting, proper, or appropriate that the stuff of a human being should undergo breakdown and subsequent recombination in the forms of sundry lesser organisms. Man clearly was meant for something more heroic than to be reprocessed through a series of plants and invertebrates, like some animal. Hence the steel-and-concrete mausoleum. Of necessity, man must persist, and in human form, for as long as there is Earth. There could be no other conceivable cosmology.

If, as I perceive it to be, species ambition is related to our morbid fear of selective pressure, then species ambition is also an important element in the continued maintenance of the conceived interspecies dominance hierarchy. If we must remain extant as a species at any cost, then no cost is too much to pay, even in the here and now, including the cost of our unilateral estrangement from nature and from life process. The double-bind is firm. The harder we struggle toward immortality, the fiercer becomes the suffocating vise of alienation. The conceptual man/nature dichotomy and the dominance structure, both spawned in our unprecedented self-conceit, are mutually reinforcing against the ultimate fear. Thus are we rendered immobile as sentient beings.

So it is that in practice – in everyday affairs – the fear is sealed off, closed out, denied. The sealants are freely offered by the humanistic tradition. Since our culture cannot admit even the faintest suggestion of the primal dread, only its substitute – the immortality imperative – remains. The conceptual surrogate becomes the reality. It follows that no new evidence, no new insight, no old memory, no heretical speculation can be admitted to question the conceptual interspecies rank-order. Human immortality depends on its safe-keeping.

It is the immortality imperative that, in an operational sense, effectively emasculates the "ecocatastrophe" and related long-term, self-interest arguments for wildlife preservation. Here again, although we can follow the conservation argument intellectually and rationally, there inevitably comes a point at which the conceptual portcullis thunders down and we once again find ourselves secure in the quiet dimness of shuttered minds. It's really quite simple. You cannot sow fear on ground that has been conditioned not to receive it. Our culture has seen to the conditioning. *What* fear?

Human immortality, which is a concept fabricated by our culture, turns out to be infinitely more potent than any argument of pure reason. Wildlife conservation, on the other hand, is *required* to be justified by pure reason, and cannot be. The fix is on.

If wildlife species are to become extinct, that will be regrettable. But any literate person knows that extinction is the way of evolution, and is in the fundamental flow of life. However, man is different. If man is not immortal, then there is no purpose or meaning in his existence. Which in turn would mean no purpose or meaning in the universe. The human immortality imperative is absolute and radical. That is why wildlife conservation has never been permitted to move to questions of ultimate value. There is no place for an ultimate nonhuman value in our western metaphysics, because *of necessity*, the human interest is the cosmic interest. That is what it is all about. Wildlife is an "externality."

The notion of necessity seems to underlie all but the most crackpot or subversive of our world views. Man is the Reasonable Being. It follows, necessarily, that the biosphere and planetary system of which he is part are explicable in reasonable terms. Since reason is based in orderly process, then it follows, as the night the day, that the cosmos is orderly. It follows still further that man is the necessary and inevitable product of that orderly cosmic process. And since man is the pinnacle of evolution, all activities that are in the human interest are by definition in the interest of the greater process.

Any observed phenomenon, whether man, animal, plant, or solar flare, or any other event, requires causes. There are reasons for all things, because all things are products of linear sequences of development toward assumed goals. It is a short hop from there to purposes, and from there to universal imperatives. If the highest purpose is the human purpose, then the purpose of the nonhuman is to serve that purpose, necessarily and inevitably. This is what we are saying every time we use the word "resource."

Causes, purposes, and imperatives are all subject to orderly classification and organization in a hierarchical system that illustrates the comparative value of a phenomenon (such as wildlife) in relation to usually undefined but assumed human ends. Certain of our assumptions, however, allow us to veer dangerously close to the final trap – the error of mistaking our own mental metaphors, symbols, and models for truth. The existence of absolute truth is of course the ultimate necessity, and the ultimate trap. I shall return to this.

None of this is to suggest that very many individual members of our species go about their daily business in constant preoccupation with matters of ultimate necessity and truth. Most of us do not. We don't have to; our most basic beliefs ("zero-order" assumptions) are taken as given. It is the unstated and unconscious assumption of these absolutes that influences virtually every human action with respect to the nonhuman; and even more important, shapes virtually every thought process having to do with the nonhuman portion of the biosphere.

The concept of necessity for our species could be illustrated in any of a number of ways, but appropriate for our purposes might be the hierarchy of human needs advanced by Abraham Maslow, since modified by others. I am taking liberties here with the original, but the sense remains. Progress through the steps is ascending, from the most fundamental (1) to the most exalted (5). In this case I would apply the scheme to humankind, not to the single individual, and as illustrative of the human necessity and the role of wildlife in meeting that necessity.

5. Self-actualization
4. Ego (self-esteem)
3. Social status ("belonging")
2. Physical security
1. Physiological base

Physiological Base

At the level of food and drink, there is little need for wild plants and animals in the human economy today, save for a very few isolated cases such as commercial fisheries, in which case the species concerned are seen not as wildlife but as a protein mine. Through the domestication of plants and animals (and of course ourselves) we have long since been freed from dependence on wild terrestrial species. Or so it is thought.

We tend to overlook the true extent of our dependence on wild nature, even today. We overlook, for example, the contributions of predatory insects in respect to agricultural "pests"; of nitrogen-fixing bacteria in the soils underlying our crops; of hawks, owls, foxes, and coyotes in rodent irruptions; of crabs and vultures and insects in sanitation; of mosquito-eating dragonflies, of water-purifying marshes, and so on. The list is not merely long; it is infinite. Nonetheless, from a conceptual point of view we are entirely emancipated from wild nature. The physiological base no longer requires wildlife, it is assumed.

If, however, some unplanned circumstance should arise to indicate that the physiological base requires the addition of new wildlife species — for whatever reason — the "knee-jerk" response is instantaneous. For example, the question of the propriety of killing dolphins in order to net tunafish can barely be articulated, much less addressed. The concept of necessity, so pervasive at so fundamental a level, precludes even the possibility of alternative perspectives, much less the formulation of questions. The role of wildlife at the human physiological support level is as basic and unquestioned as the sunrise, as unconscious as breathing. Far from being rooted in pure reason, the utilitarian role of wildlife is part of a giant autonomic nervous system that is shared by all mankind.

Wild plants and animals are *necessarily* the servants of man. How could it be otherwise?

Physical Security

Here the connection with wildlife is very tenuous indeed. Apart from those peoples who still use animal hides for clothing or shelter, the human need for security purposes quickly focuses on logs. Trees, of course, were set in place by an all-knowing providence for the purpose of supplying human dwellings. And if timber is needed — mostly, in our part of the world, for decorative purposes nowadays — that after all is what the forest is *for*. There are no doubts about that. Here, human requirements *of necessity* dictate the employment of the forest, and whether that forest contains orchids or woodland caribou or tapirs or orang-utans or indeed aboriginal people is quite incidental to the higher purpose.

In those unusual cases where a population of birds or mammals happens to be a reservoir of some disease (encephalitis, for example) that is transmissible to human beings, that reservoir must be eradicated. It would be unthinkable that there might be an alternative. My point here is not that such questions are never asked; it is that they can *not* be asked. They cannot even be framed. Questions outside the human interest do not exist, even in the most rational of minds.

Social Status ("Belonging")

Although it is clear that at least theoretically, realization of Maslow's entire ladder of needs is available to all, in practice we have difficulties at both individual and species levels. Rare, I believe, is the human individual in our society who is fully and comfortably at home in this category. It is a scarce commodity, "belonging," not in all human societies but certainly in our own.

I think that our pitiable failure here can be shown to be connected to our inability to understand the vital and radical importance of intragroup status or role in nonhuman societies. And of course our flat rejection of that biological imperative for social compliance that is ineradicably in our genes. This difficulty is impossible to overstate; it is of paramount importance, I think, in our relationships both with other human beings and with nonhuman nature. The denial of this part of our naturalness seems to me to have played an absolutely critical role in the process of our one-sided divorce, not only from nature but also from our own biology, and thus of course from our very selves.

I have pointed out that the most conspicuous and pervasive characteristic of nonhuman social organizations is compliance. Acceptance. Everyone in the social hierarchy knows his or her status or rank in relation to all the others. Notions such as status and rank are of course human concepts projected upon nonhuman systems. Here, we must drop the pejoratives and see merely "place." The notion of "rank-order" is misleading and I think destructive of our comprehension of what is going on. It is true that we can infer a hierarchical arrangement in that those close to the "top" defer to fewer and fewer other individuals, and Alpha defers to no one. What is really happening underneath all of this is that social organization is being maintained by means of unanimous compliance and co-operation, toward the general security and continuity.

You will be familiar with the ecologic model called the "food pyramid," the wide base of which contains millions of species of invertebrates, and the apex of which may represent only one species of "top carnivore," in a given locality. The pyramid represents not a reality but a process. It is a descriptive device, and a good one; it is easy to understand, and we all accept it. We seem to have difficulty, however, when we use the same device, essentially, to describe the social organization of nonhuman beings. The rabbit or the deer or the mouse or the big cat does not have a "rank" in the food pyramid, merely a *place*. Everyone does not *need* to be the final carnivore. Everyone does not *need* to be Alpha. One will do; there are other essential places to be filled.

Western culture notwithstanding, each of us does not in fact need to be Alpha. What we do need is a place, and any place will do, so long as we know it is ours. Then we belong.

Then we have completed Maslow's level three. Relatively few of us achieve this level in the fullest sense, even though the drive to find an assured place seems to be one of the most insistent in our genetic inheritance. We are denied this accomplishment in a number of ways, and the denial festers in us. It manifests itself in all of the "antisocial" individual and group behaviour from high rise mayhem to war, with which we are too familiar.

I would suggest that sheer human density is the first of the several factors that contribute to this, the most brutal of all our problems. From a biological point of view, social organization appears to be impossible to maintain in a natural way (the way in which our genes demand it) at a population in excess of perhaps twenty-five to fifty. This would be the classic "extended family," and it would seem to be appropriate to both our social primateness and our social carnivorousness. In other words, our biology. This basic need is thwarted, swamped, repressed by our absolute numbers. For it we substitute such institutionalized organizational devices as ethics, religions, laws, politics, and many more, none of which, I would advance, is "natural" to us. Whether we realize it or not, and whether we acknowledge it or not, the frustration that results is grounded in our very being.

This is an enormously complex issue of which I pretend to understand very little, but of which I suspect much. Another factor is the competitive, goal- and achievement-oriented society in which we live. I can think of no aspect of our culture so destructive to children, and thus to the society itself. Instead of raising gimlet-eyed little achievers we might try our hand at raising little social beings. I suppose there is a Richard Nixon in every four-year-old, if we foster it, but so is there a compliant, co-operative, social creature who will accept any other social creature in peaceful relationship without hangups about "rank." He will know his place, given the opportunity to flow with his own biology, and he will be content. Not, I fear, in overpopulated, overmanaged, overinstitutionalized, overcompetitive contemporary society.

So it is that with social *place* denied to us we strive instead for social *rank*, and that not being available to all, we substitute commodities, "status symbols" of all kinds, including bizarre "lifestyles," and all the other contrived and fabricated appurtenances of the craziness that is high civilization. Very few of us — an enviable, peaceful few — are able to break out of our cultural conditioning even sufficiently far as to see the problem for what it is, much less to achieve a stable place for a lifetime. (As one individual, I am only able to accomplish this prodigious feat rarely and for brief periods, but I think I can feel what freedom must be like.)

I find it deeply and sadly ironic that the most numerous large mammal on Earth, and in some ways the most successful that has ever existed, is so pathetically lonely and alien, even with respect to his own kind. The implications of all this for other, nonhuman biologic beings are obvious.

The *necessity* of striving for Alphaness in human society arises both from our biology (the frustration of the need for "place") and from our culture (ambition, competition, achievement). As I have pointed out earlier, we also have the conceptual necessity having to do with human imperialism over the nonhuman.

As a species, our sense of belonging in nature, our sense of a place in nature, has been utterly destroyed. The concepts of "man/nature," the drive toward interspecies dominance, and the human immortality imperative have seen to that. This is the unilateral divorce from life and living that is the unique accomplishment of our civilization.

Having wilfully abdicated our place in life process, we can

no longer remember that "place" means "belonging," and that belonging is what living is all about. Since we can no longer conceive of a natural system that includes us, we feel estranged, alien, even resentful. Our bewilderment can turn ugly. This of course *is* the necessary outcome of our conscious and deliberate denial of nature, our nature, our life sphere. From this springs the paranoia that is the hallmark of every thought and action with respect to those aspects of non-human nature (weather, disease, death, rose aphids) not yet brought under our absolute control. We remain at level two.

Ego (Self-esteem)

When it comes to self-esteem, the alien is in bad shape. Solitude can weigh heavily on the self-image. Nothing can exist without a context; the context in which we now live, however, is one of our own imaginative creation. In spite of our posturing, it does not and cannot satisfy us as sentient beings. So it is that the self-image of mankind is in parlous condition, notwithstanding our protestations to the contrary. Our defensive attitude and actions toward nature are ample evidence of that, as is our problem with rank.

To me, it all seems very simple. So long as we are *necessarily* qualitatively different from, superior to, and exempt from natural processes – from our own being – the paranoia will necessarily follow. For the technosociety that cannot find a place at level three, self-esteem lies beyond comprehension.

Self-actualization

This highest level in the hierarchy of human needs is sometimes called "self-realization." Whatever either term really means, one would assume that is a purely individual affair. If it means becoming the fullest being of which one is capable, having in mind both nature and nurture, then you could say that in the absence of a standard the notion is meaningless or impossible, or you could say that by the mere fact of having lived, everyone achieved just that – everything of which he was capable. I do not know that. But I think I know that total achievement of the human "potential" (whatever that may be), whether for the individual or the collective, will not be possible outside our biologic context.

It is assumed in the Maslowian hierarchy that one must move through the levels in sequence, beginning with the physiological essentials, through physical security, and so on. If the abandonment of our own life context will not permit us so much as to reach a state of "belongingness" and thus of "self-esteeming," then it would seem that actualization for the alien species would be supremely difficult. We would need, for the process to complete itself, a relationship with the nonhuman which our culture cannot accept.

My own interpretation of "self-actualization" is death and recycling. I think the supreme achievement would be conscious abandonment of "self," and the sensation of release from life-long mental slavery, the joyful feeling of rejoining that very old and yet immediate familiar, the all-encompassing and all-receiving life process. At this point one knows "compliance." Having found it, one knows at last what had been forgotten, and what every nonhuman being knows with its every breath and every pulse, and in every cell.

Self-actualization for our species is, I fear, distant. It would require our acknowledgement of belonging in nature, our development of actual (as opposed to artificially conditioned) self-respect as sentient beings, and it would require open acceptance from the inside out. Not, I fear, just yet.

Our failure to see ourselves as and thus to again become an integrated functioning species in a dynamic context, and our inability to move past that point, arise entirely – it seems to me – out of the concept of human necessity. None of our most bizarre conceptual fabrications has had such far-reaching consequences for us, and thus for all of nature.

Our cultural tradition – the set of "givens" which engender, nourish, and shape both our attitudes toward nature and our overt actions with respect to nature – is a complex of concepts so deeply held that in practice it is rare even to acknowledge their existence much less to attempt to evaluate their effects. Here I have called attention to only a very few of them, but they are among those that I believe to be at the root of our difficulties in entertaining the very *notion* of wildlife preservation, much less implementing it in practice.

The conceptual qualitative separation from nature, the imagined human dominance over the nonhuman, the imperative immortality of the human species, and the notion of necessary human directions, goals, purposes, and reasons in the universe are of bedrock importance to all that grows out of them. They dictate not only how we frame the argument for wildlife preservation, but also how we fundamentally *perceive* wildlife, ourselves, and life process.

THE PERCEPTUAL PROBLEMS

Our concepts – our ideas, thoughts, notions – about ourselves and the world around us both create and are created by our perceptions. As human beings, which is to say, primates, our sensory apparatus is heavily biased toward the visual. Thus we have "views" about ourselves and the biosphere; we "see" these phenomena in certain ways. Were we dolphins, no doubt we would have "sounds" or "echoes" about things; were we moths, we would forever be trying to make "scents" of things.

The handful of concepts I chose for attention in the preceding section all have close relationships with our perceptual activity. In fact, those relationships are as intimate as those in any ecosystem, with feedback mechanisms so complex and subtle that they relegate the chicken-and-egg problem to primer level. So interwoven are the dynamics of perceiving and conceiving that simple cause-effect labels and descriptions are worse than useless. They are downright misleading. In full knowledge of this, however, we continue in our stubborn positing of universal causes and effects, laws, reasons, beginnings and ends, one-to-one relationships, special status, and the mechanical explicability of all things. I often think that the most noteworthy characteristic of the Reasoning Being is his immunity to reasonableness.

This is not an essay in perception, but I must indicate my ground rules before proceeding to our perceptual problems with wildlife and its preservation. Perception has many definitions; the one I shall use here is "a hypothesis of the world based on sensory information." We are forever testing our hypotheses, matching internal "realities" against external ones, testing our inner worlds, which are always changing, against the "real" world. I see this constant process of change (growth, development) as necessarily including the process of identification — sorting out and identifying the sources of information that we receive through the five standard sets of sensory apparatus. We smell steak, we hear thunder, we feel sunburn, we taste whiskey-soda, we see the sunset. And of course we can do all of them at once. All of these sensory experiences are perceptions.

Always, in perception, there is much *imaging* going on. There is much picturing. As a continuing process of hypothesis testing, including identification, perceiving involves the *matching* of external stimuli with some internal image. If we did not know what we were sensing, at least most of the time, we would not be able to function at all. If I say "Red," you will instantly produce an internal image of red, but probably not the one I have in mind. If I now say "Rose-red," your internal image will be closer to mine.

Most of the time this process of internal imaging is the next thing to instantaneous. What we actually seem to be doing is *testing* a sensory impression (sight, sound) against a stored inventory of prior experience of that impression, which results in a match, an identification. It is like looking up the bird or the flower in your field guide. If, when you are walking through the apartment corridor, you don't happen to have on file the sensory experience of corned beef and cabbage, you cannot make a match, and you won't know what it is that so massively surrounds you. Or, if the experience in store doesn't *quite* match the smell, you can always force-fit pigs' hocks and sauerkraut. Force-fitting is a major perceptual problem.

I have a little image of my own that helps me to understand this process of perceiving. Picture inside your head one of those circular wheel-like file-card holders that you can rotate. You spin it around until you find the proper place, then stop it and read the appropriate card. It's like that. Each of our heads contains one of these devices, filled with cards, spinning around perpetually. When we receive a sensory signal of some kind, the wheel stops spinning, and up pops a little bubblegum card – what I call an image-card. That's what happened when I said "Red" and when I said, "Rose-red." No library cross-referencing system was ever so efficient.

We have to remember that one-to-one matches are not the rule. It's more complicated than that; no library has just one entry under each heading. There are many. So it is with our own filing systems. Sometimes it may take half a dozen or more cards to make even an approximate match. But we do the best we can.

Seen in this way, each image match is an event. Although perception is a constant and continuing process, it is made up of a series of countless discrete little matches, all of them part of the greater sensory perception process. Each time we make a match, the wheel stops (the world does not, and that is another problem). An individual perception is an event. "Red" is history; now you perceive "Blue." It is very much like a motion picture film, a series of individual frames. Each individual frame, in isolation, is soft and poorly defined, as the result of having been exposed while moving, and at slow shutter speed. A single frame from a piece of film is not satisfactory viewing. But strung together and projected at proper speed, that series of quite poor photographs at once becomes sharp and crisp and focused – the greater context. All of which is a way of saying that in spite of reductionist science, the fragments give precious few clues as to the nature of the whole.

In this world, at least, all things seem to be finite. For most of us, there appears to be a limit to the number of image-cards that we can accommodate in our little whirligigs at any one time. There can be an immense number of them – and there is – but there is still a limit. Not an absolute limit in a biological sense, but a limit to what our culture is capable of imprinting upon us. It follows from this that everything – literally everything – hinges on what image-cards happen to be available at a given moment. Granted, we do shuffle them constantly as the consequence of simply being awake and moving; we sort and re-sort, tossing out old and inserting new, but at any instant we are limited in our perception to whatever image-cards happen to be in inventory at that instant.

Naturally, a problem emerges straightaway. Suppose whatever happens to be on file turns out to be out-of-date, or printed off-register, or upside-down, or badly translated, or wrongly filed — or indeed there is nothing there at all with which to match the next sense impression that comes in? You can see how some spectacular mis-identifications can occur. Misinterpretations. Misunderstandings. But of course — and here is the *real* problem — we have no alternative but to believe these matches to be accurate. Seeing is believing.

I happen to be a birdwatcher, and all birdwatchers treasure their skill, knowledge, and their virtuosity. In fact, many of us compete overtly and outrageously in the swiftness of our card sorting and the size of our inventories. Suppose I am sitting at my desk, and some faint bird song drifts down to me from the big shade trees outside. Now, I have stashed away a goodly number of image-cards representing birds of which I have had prior experience or I have copied from the texts. First, some mechanism that is so swift as to appear automatic, sorts the file geographically and comes up with the batch for southern Ontario. Then I proceed to match the sensory impression (sound) against that local batch, and out of the pile falls a bubblegum card, in full colour, of an American robin. I perceive an American robin, having had an image of one with which to match my sensory impression. Not any other kind of robin, of which there are many, but *that* one.

But, God help me (and my reputation), what if the filing system comes up with an image not of a robin but of something else; something that sounds a bit like a robin, but is not one? Perhaps a rose-breasted grosbeak. In that case, I'm in trouble, and my bird identification will be wrong, but I won't *know* that it's wrong. So how can I go about distinguishing between an accurate picture and an inaccurate one? They are both real, to me, and that is all that matters. They are both images, both in inventory. How do I choose?

Naturally there is always the possibility that I may have no image-card at all with which to even try to make a match. This happens regularly when I travel in unfamiliar places. In such cases, I simply will never know, and I will have missed a particular experience. This is one of our most radical perceptual problems. We are stuck with the image-cards we have – and only those – but sensory impressions are continuous and all-pervasive; we cannot shut them out. All we can do is *try* to make a match, and the margin for error is high. In fact, if you allow yourself to brood about it for a while, this potential for error becomes quite alarming.

Any aggregate of people – a cult, a community, a society, a culture – has a collection of image-cards that most if not all of its members share in common. When I say "Santa Claus" most North American readers will retrieve essentially similar image-cards. An Inuit or a bushman or a Chinese draws a blank. Most Europeans would display consistent images, but they would be a bit different from the North American ones. On either side of the Atlantic, however, there would be indignation and outrage were Santa to appear in a Batman suit. There is a vested interest in our images, and it is strong. I would refer you only to the image of human dominance over nature, for one example.

Each society has an inventory of images that is shared by nearly everyone. In each society, we have not only shared trees and colours and birds and folk heroes, but also we have shared ethics, theologies, philosophies, science, imaginative literature, music, graphic arts, aesthetics, beliefs, and assumptions. Remember that these image-cards are put in place by experience. Societies have experiences such as Santa Claus that are shared right across the board. Such too as nonhuman nature.

Another dimension to all of this is that while we share culturally processed images we also share a common store of *probabilities* about the likelihood of such-and-such a thing being the case in such-and-such a set of circumstances. If the man far down below me in the street has got his umbrella up, then it is probably raining. If I see a little fluttering group of ivory gulls at the edge of the arctic ice-pack, there is probably a polar bear around. And so on. I will *probably* get my bird song right after all, because I have in reserve an approximation of the odds on the presence of various bird species around my house.

Notions of probability can be set in place either by personal experience or by cultural conditioning – usually a bit of both. I can read the geographic distributions of the birds in a book or several books, or I can learn them the hard way, by working them out for myself, in the field. Generally I will tend to accept the literature, especially in some "new" environment.

Our notions of probability are almost entirely subjective (we very rarely go around objectively estimating the odds on the sunrise, for example). We accept them as given. Givens are *cultural* image-cards, and it is against these that I weigh, sift, filter, and attempt to match the various sensory impressions from the outside to which I am exposed continually. It is through these givens that I perceive. My culture fits me with a set of pre-ground goggles through which to receive the world.

To be alive is to be experiencing a perpetual barrage of sensory impressions – sights, sounds, smells, and all the rest of it. In fact, in order to make living bearable, we have the facility to screen out or ignore much of that barrage from minute to minute, and to concentrate on particular portions of what is available. You screen out the barking dog, the screaming infant, the overflying aircraft, and at least occasionally the voices of those nearest and dearest. Your eyes can concentrate on the bird while dropping the bush into limbo. You can tune in to the voice of a conversational partner no matter how many conversations may be hububbing around you. You will pick up the scent of smoke no matter how many other odours may be swirling about the kitchen. This is a matter of some in-built capacity to select our impressions and also to lessen the potentially distracting or even maddening effect were all receptors to be left open to their fullest extent at all times. It's an unconscious protective mechanism.

There are times, of course, when we consciously reduce the barrage of sensory stimulation, as when in self-defence we take the telephone off the hook, turn off the radio, shut the window and door and try to get on with a backlog of homework. Then, we are consciously simplifying the sensory environment in order to concentrate as fully as possible on the task at hand.

There are still other times, however, when our sensory impressions are *severely* diminished, as for example when we enter virtually any contemporary institutional building, office complex or apartment high rise. Then we encounter the phenomenon known as sensory deprivation. In sensory deprivation, we are told, the victim experiences not lessened but *increased* imagery. He does not however experience perception as I am interpreting it here. Apparently what happens is this: the increased internal imagery is the result of the perceptual apparatus trying to keep going in spite of reduced or almost totally absent external stimulation. The little rotating wheel keeps spinning madly along, throwing up one image-card after another – including, it is said, many rarely used in usual circumstances – as though trying to find *some* external stimulus with which to make a match. It throws up card after card after card, wanting and trying in every conceivable way to keep in touch with the outside. But there is nothing coming in, nothing to match with, and the internal images are therefore meaningless. There is nothing. And of course we can never have an image-card with which to match nothingness because we have never experienced nothingness. We cannot experience nothingness because our culture cannot *accept* nothingness. So the wheel of sensory misfortune spins, and keeps spinning.

It is at this point that I advance my thesis with respect to the sensorily deprived society. I believe we live in a society of sensory deprivation, with all of the bizarre imaging that that implies. Now, whenever I put this notion forward in a lecture or some other forum, I am immediately pounced upon by the urbanists, who gleefully point out that I am obviously alone in an echo chamber of my own creation, because everybody knows about the painfully obvious urban phenomenon known as sensory *overload*.

Everyone knows that, far from being sensorily deprived, the average city dweller is fairly overwhelmed by a shattering and inescapable barrage of sense impressions – flashing signs, screaming horns, choking exhausts, bellowing shills, and all the rest of it – kaleidoscopes and cacophonies of every conceivable kind. Bumping, jostling, sweating, swearing humanity. And all of that is true. Certainly, in the city we are grievously and continually assaulted, and we defend ourselves by screening out that which we cannot or do not wish to assimilate.

But there is more to it than that. I believe that this urban sensory assault and battery, if we feel we must describe it as overload, is a simple quantitative overload, having no variety whatever. The tumult and the sensory hammering all come from one single source; without exception they are of human manufacture. All of our sensory bruises are self-inflicted. And when at last the tumult and the bombardment abate, we are left with the only monotonous landscape on Earth –

the human landscape. Monotony is a prime contributor to the symptoms of sensory deprivation.

In a qualitative sense, far from being sensorily overwhelmed, we are virtually starving for variety. Something, anything, that is not of human manufacture, human fabrication. I can conceive of no other situation in which sensate beings are harshly and uninterruptedly imprisoned in this way, exposed entirely to sensory stimuli from a single source – themselves.

This is very sad, and I think very serious, because the problem feeds and compounds upon itself. In sheer physical retreat from the constant, terrible insults to our delicate sense organs, we tend to narrow even further the range of stimuli that we allow ourselves to receive. We find ourselves walking about in an even narrower prison of our own personal making, in sheer self-defence. Thus, whilst reducing the absolute quantity of the barrage, we reduce still further the range of potential variety. We reduce still further our own access to possible qualitative experience.

As indicated earlier, I see "quality of life" as the care and maintenance of options. I believe that *sensory* options — textures, scents, tastes, pictures, sounds – are the most fundamental and necessary of options, and those that we most consistently and perversely deny ourselves. This, essentially, is what I mean by sensory deprivation in civilized society.

I am trying very consciously not to be "romantic" about all this. I am not, by the way, a subscriber to the "wilderness mystique" and all of that. I mean, quite simply, that we very deliberately deny ourselves sensory access to the radical nature of being.

Thus it is that we find ourselves estranged not only from the life context external to ourselves, but also from our own being. *We are all alone.* No wonder we behave the way we do.

Strange things happen in solitary confinement, and strange things happen too in high-density confinement. Paradoxically, the individual finds himself in high-density "solitary."

Incidentally, like others who have thought about this problem, I am surprised not so much by the high level of crime and other madnesses in our cities, but rather by the relatively *low* level of it, having in mind the appallingly deprived and degraded circumstances in which, as sentient beings, most of us live. From this point of view, I think, mankind may be even more admirable than the most zealously dedicated of the urban humanists have yet imagined. I can think of no other large mammal that would tolerate it. The blood would run freely in the gutters *every* night – not just on week-ends.

I think also that the extraordinary boom in the house plant business these days is a very good indication of a fundamental urban problem. I do not see it so much as a decorative fad as the expression of a deep and primal need – a biological imperative. To be alive means to be sensate means to be *in touch*. We have to maintain contact with something – anything – that is *alive*. It's the most fundamental part of being. The geranium on the tenement windowsill is both an offering to the mysterious tidal pull of some distant biological memory, and a heartbreaking cry for help.

Assuming as I do that this quantitative sensory barrage coupled with qualitative sensory undernutrition adds up to sensory deprivation of a very real kind, some of our most peculiar behaviour seems to make a little more sense. Like that of the experimental volunteer shut up in the soundless empty room, our internal perceptual machinery keeps throwing up its little images in its search for matching sensory impressions that are not there. It follows that as a society or a culture we experience a series of quite meaningless pictures:

hallucinations, illusions. This I believe to be the common perceptual experience of the sensorily deprived society.

This society is of course the spawn of its environment (the interaction of its culture and itself). Our present predicament has entirely cultural roots. The tradition out of which we have come has, by a process of artificial selection as sophisticated and complex as that of any geneticist, produced a cultivar so remote from its origins that they are barely discernible. We have been unceremoniously ripped from our life context, sprayed in plastic, and packaged in hermetically sealed containers. Sealed off from our own aliveness, including our sensory processes. All this has been done through cultural conditioning. We have been divorced for so long not only from biophysical nature but also from the very *awareness* of biophysical nature that our images of the world and even of ourselves have become little more than institutionalized hallucinations.

Our culture has institutionalized such hallucinatory images as human supremacy over all animate and inanimate nature; a swollen, obese GNP as the very model of social health; high technology as the Holy Grail, and so on. There is even a picture of a planned, organized, ordered, and thus understandable and manageable universe. The trouble with all of these counterfeit bubblegum cards is that there is no external "real world" opposite number with which to match them. They are displayed into emptiness; they are without meaning. They are, in my view, precisely the sorts of hallucinations that arise in sensory deprivation, in the absence of natural sensory stimuli. If perceiving really is hypothesis testing, then many of our most treasured hypotheses are not testing out.

So it is, I think, that our perceptions (and our concepts) having to do with man and nature and the cosmos, are images that reflect the "real thing" about as accurately as toy poodles with painted claws resemble the wolf, as assembly-line leghorns resemble the red jungle fowl, as glazed castrated Herefords swimming in drugs resemble the great white aurochs. The image no longer matches the reality. Our culture sees to it that sensory impressions from the natural world are consistently screened from us, so that both as individuals and as societies we are permitted to perceive only distorted images of our own queer, tortured, internal fabrication.

In this sense our hallucinations about the relationship between man and non-man, for example, become understandable. By closing off our connections with external reality, which happens to be our root system, our culture drives us to dependence on entirely artificial support. We are left with a jumbled catch-bag of monstrous illusions, and what is more, we take our hallucinations for truth. Seeing is believing. Even more important, believing is seeing.

The gestalt psychologists explain many of our problems in terms of what they call "closure" — sealing off potential sensory impressions (or ideas) which might otherwise rock the boat, simply by penetrating the human consciousness. This must be avoided at any cost. After all, my personal set of institutionalized beliefs is complete and finite. So is my society's. Don't intrude. Don't annoy me with facts. Closure is described as the attempt to see all things as neat and orderly and symmetrical, and to ignore holes or anomalies. We want to see everything (including ourselves) nicely and evenly balanced; we want an idealized perfection. We feel secure that way. (Professional proof-reading is often given as an example of what closure is *not*. Most of us make terrible proof-readers because we read for content, and close out the typographical errors. We never see them.)

But any naturalist should know that there is no security, nor any beauty, in symmetrical, balanced order. The bird-watcher looks for differences, not similarities. Two things the conceptual ecologic model does not demonstrate are symmetry and balance. If it were to exist, symmetry would indeed be fearful. The hoary old "balance of nature" image-card does not imply a static, controlled state of affairs. The so-called balance swings and yaws and bumps and pitches quite asymmetrically around an hypothesized homeostatic line. But homeostasis is never achieved. If it were, everything would thereupon lock and stall. Stalling is not what life is all about.

Unfortunately, our western institutionalized perceiving seems to be predisposed to cleave to the hallucinatory image of balanced symmetry, which is a life contradiction. Nonetheless, the drive is very real, a kind of cultural imperative. This appears to be manifest in the urge to reduce perceived "chaos" or disorder (which is bad) to manageable and controllable order (which is good). Unfortunately for all of us living sentient beings, the human assumption of the possibility — no, the necessity — of perfect balance and symmetry, whether in the individual, the society, the landscape, the biosphere, or the cosmos, looms very large as a potentially disastrous self-delusion. Such a belief, when force-fitted to outside stimuli, can result in some very strange misinterpretations, which in turn may be seen as indivisible and absolute cosmic truths.

Predisposition is exemplified in "optical illusion" drawings, especially those at which you may have three alternative ways of looking — a choice of perspectives. You can see only one perspective at a time, because that is all your brain can handle. Once you have made your first impression, however, you tend to "lock" into it, and it can be difficult to switch off that (acquired) predisposition to see the same picture from a new perspective. This is the way, also, with cultural predisposition.

From the point of view of wildlife preservation, our western perceptual predispositions are clearly and sadly illustrated. Such is our cultural image-making apparatus that it is the next thing to impossible for us even to *perceive* wild plants and animals as anything other than human amenities, utilities, or commodities. The humanistic tradition has done its work so well that it is the next thing to impossible to perceive any arrangement other than a man-centred and man-dominated universe, any purpose other than the human purpose, any process that was not designed for the exclusive benefit of man. It is next to impossible for anyone, however highly motivated, to see wild plants and animals as more than man's wards, "incompetents" in our custodianship.

Thus it is that the less well-motivated, and the less-informed, see natural predation as murder, tunafish as a gold mine, dead porpoises as externalities, human mortality as unnatural, wild nature as alien and hostile enemy. The roots are long, and deep. They nourish and produce not only our arguments for wildlife conservation, not merely our assumptions, not merely our concepts, but also our very perceptions of that life tapestry of which man is but one part in millions.

I believe that we are what we perceive, no more – and no less. This is not meant to be a value statement; it is a descriptive statement. I have been a naturalist and preservationist for over forty years. If you are not a naturalist, then I have perceived more widely than you have, and I have experienced more widely than you have. (This is not meant to sound presumptuous; if your experience, however rich, has been entirely in the human context, then mine has been richer.) I do not attach any "value" to this in a comparative way, because the only value has been for *me*. I know that I perceive differently than the non-naturalist perceives not merely in knowing the labels for the biophysical surrounding and its parts, and in being able to understand some of its process. The fundamental difference is that I *feel* (sense, apprehend, conceive, perceive, experience) them differently than the non-naturalist does.

This is not an attempt to turn you into a naturalist if you are not already one. To be perfectly frank, I have no emotional investment whatever in the expansion or the constriction of your experience. I am no more than human; I don't know you and I cannot "relate" to you or to your emotional or intellectual welfare. What I *do* care about, of course, is the way in which your perceptions of yourself and of nature may influence your actions with respect to wildlife preservation. And that's *all* I care about. But in my universe, that's everything.

And so it becomes blindingly apparent to me that it is in my interest to talk about your interest. If you were to become a naturalist, I would be the beneficiary. Your becoming a naturalist would further my investment in wildlife preservation. This argument is not addressed to your self-interest in wildlife (whether for altruistic or consumptive reasons); nor even, as an argument, to the interest of wildlife as such. It is in *my* interest.

This is where we join hands. There is no interest but mine – and yours. There is no *reason* for you to become a naturalist or wildlife preservationist, except for your entirely selfish individual experience, your intellectual-emotional experience. Any other argument, I believe, is spurious. By adhering to conventional (rational) arguments for wildlife preservation, and by continuing to obey conventional rules of argument, we illustrate the limitations of our own perception and of our own individual and collective experience.

·∽◉◡· FOUR ◡◉∽·

EXPERIENCING

If, to this stage, this essay has appeared to be a deliberate attempt at vivisection, I suppose that perception would be justified. I have tried, as best I could, to strip away at least a few of the more obvious layers of rhetoric that surround the corpus of wildlife conservation arguments, and to de-mystify that rhetoric toward the revealing of what seem to me to be the essential issues. Of course when you set about unwinding the rhetorical cocoon from almost any concern, you run the risk of exposing false expectations for what they are, and of experiencing at least some measure of disappointment. Shocking to relate, it frequently appears in the final analysis that there was not very much there after all. This is not an appealing notion, but it does serve to remind us why the rhetoric was there in the first place.

The argument for wildlife conservation has been variously framed and in various species of rhetoric because it was felt necessary to present it in those forms in the absence of a cultural belief system that could accommodate wildlife preservation – in my terminology – for its own sake. That, in practice, would clearly be a preposterous argument. Ergo, suitable rationalizations had to be found, and some of these we have discussed. For reasons I have already outlined, I believe that these rationalizations are incapable of meeting the needs of wildlife. Our uniquely human self-deluding is practised even in endeavours such as this.

On the basis of some experience in conservation affairs, I am at last persuaded that mere argument as such is entirely worthless. Indeed, as I have tried to show, some of our arguments (certainly the most-used ones) may even be detrimental to our purpose in that they tend to reinforce the very belief system that gives rise to our problems in the first place (evidence, for example). Argument, it seems to me, is never going to help wildlife. It rarely has, and there is little to persuade me that it ever will, appreciably. If anything is going to help wildlife it is individual people. Not people in the mass; after all, the public is no more than a concept. The reality is an aggregate of individuals. I believe that wildlife preservation is entirely dependent upon *individual human experience.* Unashamedly, I shall now set about attempting to build a rhetorical environment for that belief.

I have insinuated the notion of "experience" at various points. Individuals such as myself, who for various reasons have been unusually fortunate in having enjoyed a wealth of experiences with wildlife in many parts of the world, eventually are forced to adopt a kind of self-protective defensiveness about those experiences. Who are you, someone will say, to recommend cutting down on the flow of tourists in Africa, in the Pacific Islands, in Indo-China, in South America, in the arctic? You've been there; you've seen it. It's all right for you to say these things. Yes, of course, I am supposed to say: *mea culpa.* Yes, I have been in all of these places and more, and yes,

I say there are too many of me. What else can I possibly say? Yes, I recommend denying those experiences to you — in the interest of wildlife.

First of all, I am not in the least embarrassed by or apologetic for my good fortune and what may appear to be your lack of it. I shall attempt to show that all is available to you anyway. All of it. That is because the "nature" experience is entirely qualitative, not measurable, not rational. You will have to accept on faith my statement that nothing has been added to my qualitative experience since I was ten years old. There have been many quantitative overlays, many adventures, many data, many sights and sounds, and all the rest of it, but the quality has remained absolutely unchanged since my childhood. The initial experience was everything, and has continued in its original form life-long. Without it, subsequent adventures would have meant no more than those of any other tourist.

There was behind my parents' house a city ravine, with a little stream running through it. At one end, before the stream disappeared into a large pipe, there was a little marshy area where the water spilled shallowly to one side. There, there were toads and frogs and newts. If you lay very quietly in the grass at the edge, you could observe them. The longer you looked, the more deeply you were mesmerized . . . possessed. There was no world whatever, outside that world . . . nothing beyond shimmering light on water, smooth clean muck, green plants, trickling sounds, flickering tadpoles, living, *being*. That was when the pain started.

The knife of separation is cruel. I not only remember in a factual sense but I can *feel* to this day the anguished frustration, the knowledge that I could never — not ever — be more than a boy on the grass, excluded from that world wholly and eternally. But why? Why pick on me? I wished it no harm;

I only wanted to be part, to join, to "plug in." The denial was impersonal and cold and final. It has gnawed at me ever since – not all the time, mercifully – but much of it.

I wept over it, in a dogwood thicket. In the certainty that through no apparent fault of my own I was being unjustly denied something that was as fundamentally important as air, I felt much anguish at times. Unpredictably, of course, as it is with preadolescents, there would be unexpected moments of pure and inexpressible joy and happiness when the "free flow" between nature and myself was unobstructed and open. Such moments always seemed to happen accidentally: why couldn't I *will* them? Always there was a mix of sadness and pleasure. My early experience with nature was bittersweet; it still is. I rejoice in wildlife and I despair, in equal measure.

That is one side of it. Plans were revealed for the construction of a storm sewer through "my" ravine. Shock, dismay, and all the rest of it were mine early. The ten-year-old mind is not subtle: how can I warn the frogs and toads and newts? Can I get them out of there, take them away somewhere? They are defenceless; it is wrong to hurt them. What right do we have to hurt them when we cannot warn them? They don't know what is happening, or why. There was much puzzlement here. All logic seemed to be backwards or upside-down; nothing made sense. I could do nothing but watch, with sorrow and fury.

But why the sorrow and the fury? What *is* compassion, after all, and where does it come from? And why do so many other people feel nothing at all? Those questions are as germane today as they were when I was ten. It seems clear now that, although there was no gainsaying the intensity of my emotion, my feeling for the wildlife beings involved, the sorrow and fury, were perhaps entirely on my own behalf. I was responding intensely because *I* was being impinged upon.

I think that through those moments of "free flow," in the grass by the pond beneath the dogwoods, the toads and the frogs and the newts and their hypnotic sunlight had been irreversibly incorporated into my world, literally into *me*. *My* world was being tampered with; *I* was being invaded. Next spring I would have a piece missing, chewed out of me by the ditch diggers. The hurt was much more than resentment and sympathy. It was real, and I would feel it always.

It can never be otherwise. You, the individual experiencer, are paramount, because the emotional investment is yours alone. There is but one universe and it is you. But the experience need not be immediate and on-site. Despite repeated attempts, and despite even having heard and smelled him, and having photographed his fresh footprint, I have never *seen* a wild tiger. At this late date — meaning both my own chronology and the status of the tiger population — I probably never will. That is not as important to me now as it used to be, because the fortunes of the tiger are no less close to me for that. The tiger is already an integral part of me, and his fate is mine. He entered me with the toads and frogs and newts. My experience is *my* experience and the physical proximity of the tiger really has rather little to do with it. I shall be just as depressed by bad news of the tiger, and just as elated by good news of him, as I would have been had we lived side by side. Wildlife preservation "for its own sake" is for the sake of the experiencer. Each experience, whether physical or intuitive, adds to my universe. Each extinction subtracts from it.

Now, this is not an argument of self-interest. Earlier in this essay I decried those arguments because I believe them to be inherently weak at best, dishonest and destructive at worst. Self-interest is just that: personal advantage, personal motive, personal gain, personal profit. That is not what wildlife preservation, in the "ideal" sense, is. Wildlife preservation for its own sake – for your sake – is none of these things, because it cannot be viewed comparatively and it cannot be measured. It is an experiential phenomenon only, and it cannot be "valued" (weighed). It happens singularly and uniquely and exclusively in your universe. Wildlife as part of your world, and part of you, and that is all that matters. That is what preservation is.

I have no doubt that it will be charged that this is not a reasonable argument. It is too "mystical," not sufficiently logical. I feel that the conventional arguments for conservation are unequal to the task, not because they are not rational (most of them try to be), but because *there is no rational argument for wildlife preservation*. There is no rational argument for personal, individual, qualitative experience. There can be no rational argument for your universe or for mine. They simply are – whatever they are. Neither you nor I need accept any obligation to rationalize those universes; indeed such exercises have a way of being severely self-defeating.

My point here is quite simple. Rationalizations are substitutes for qualitative (as opposed to abstract) experience. Once the experience has been incorporated into oneself, there is no need for either rationalization or proselytism. There is no "reason" for wildlife preservation. It is a state of being.

There was a time when sensible people knew that reason and experience could not be torn asunder; together, they were "knowledge." It is this knowledge of which I speak – the state of being that is wildlife preservation. Today, however, its two parts are conceptually polarized. We have rationality (science) and we have mysticism (non-rational experience). We have intellect as distinct from emotion. We have reason/feeling, and we have man/nature.

As indicated at the outset, I have complained long and widely about the intellect/emotion duality in the conservation context. Anyone with a shot glass of intelligence can follow most logical argument, but inevitably there appears some "block" that prevents the *emotional acceptance* of man as a biologic being, with all that that implies. We have forgotten that what makes the world go round is compliance, not logic.

The potentially fertile union of reason and non-rational experience is made impossible for wildlife conservation because there is no emotional impetus for that union, but (in our culture) much resistance to it. Remember that the "academic" problem is one in which there is little emotional investment in the solution but very much in the method. Wildlife preservation, as I understand it, is the antithesis of the academic problem.

At the root of our wildlife preservation difficulties are the perceived dichotomies that flow from the "mind/matter," "spirit/flesh" dualism inherent in western cosmologies. And not only are mind and spirit separate from matter and flesh, they are *superior* to them. Mind *over* matter, remember. So it is that man stands to non-man as intellect stands to emotion. Man/nature: intellect/emotion. There you have the fundamental and operational formula of the technocratic age and the failure of wildlife preservation in that age.

It seems to me that just as our culture teaches us to perceive a gap between intellect and emotion, at the same time we feel some "irrational" (inherent, natural) urge to seal over or close that gap. But we are prevented from doing so. Culture covers the institutionalized gap with an artificial membrane, a fabricated gloss of beliefs, givens, religions, philosophies, ideologies – rationalizations. Remember that rationalizations are substitutes for experience. For purposes of the scientific-technocratic juggernaut it is necessary for us to *perceive* a gap between intellect and emotion, but that gap must on no account be experienced in a qualitative sense. It would vanish forthwith – and then where would high civilization be? So it is covered over with a transparent plasticized veneer of assumptions so that it may be seen but not touched.

So it is with the necessary gap between man and nature. It must be maintained at any cost, for reasons I have already put forward. Unlike the gulf between intellect and emotion, however, it *must* be "experienced" (in pre-selected fragments) for its continued maintenance. Here the *illusion* of experience is provided through the same beliefs, givens, religions, philosophies, ideologies – the same panoply of rationalizations. Here again rationalization is substituted for the whole (idea + quality) experience, because of course the man/nature gap cannot be experienced. It doesn't exist. Man, after all, *is* nature.

The intellect/emotion gap must not be "experienced" because if it were, the experiencer would *know* (reason + experience) that it does not exist either. The ordered universe of the Rational Being would fly apart. In the final accounting – or the ideal knowledge – there is of course no abyss: man and nature are one, intellect and emotion are one. I experienced this – although I could not articulate it – at ten years of age. Both dichotomies are entirely conceptual and artificial. Both are purely symbolic.

As I have pointed out earlier, we have a nasty and persistent way of taking our symbols (metaphors, models) for absolutes – for truth. Thus we accept the "reality" of order and purpose, beginnings and ends, predictability and control, meaning and direction in the anthropocentric universe. Ideas, all of them. Rationalizations that cannot be corroborated by experience. In the classical sense at least, scarcely "knowledge."

Symbolic representations are used for many reasons,

including as a kind of shorthand. Perversely, however, we tend to confuse our shorthand for the real thing. Thus "God" becomes infinitely more real than that which it represents; "species" becomes the reality as opposed to the process it symbolizes; "hour" is seen as concrete and tangible, no longer a conceptual unit of measurement. Usually, apart from such as mathematics and music, where the shorthand is a genuine tool, symbols turn out to be no more than metaphors for that which is inexpressible – most often for that which is not understandable or manageable. And, as often as not, symbols are used to metaphorically conceal root fears. The rationalization is substituted for the experience out of fear of the experience. Thus, when I was very young, flags and bands used to ennoble war. The vivisectionist never kills an animal; he sacrifices it. You never perform autopsies on people, only on bodies.

For me, the most interesting symbolic-metaphoric rationalization of all time is the uniquely human immortal soul. In my earlier discussion of the immortality imperative, I remained at the species level. There is no *individual* immortality imperative. We know better than that, on purely empirical grounds. So we invent a comfortable continuity called the afterlife, together with a vehicle for its achievement. No objective evidence can ever shatter this one; no contrary experience can ever be brought to bear. It is unspeakably gorgeous; of all the products of human ingenuity it is supreme, for it can never be disproven. Here, it seems to me, the institutionalized intellect/emotion dichotomy is revealed for what it is.

I now see the root fears at work in these issues somewhat differently than I once did. In western post-renaissance tradition there is a fundamental vested interest in the preservation of reason from the inroads of feeling (man/nature).

The buffer between them is of cultural fabrication, a rational response to an irrational fear. So it is with individuals, well and truly conditioned as we are. The fundamental fear, I think now, is of disorganization – of emotion eroding the integrity of the conceptual (reasonable) edifice that, we believe, is each of us.

Remember the classic accusation that is levelled against wildlife preservationists. We are *emotional*. When confrontation with our opponents reaches the abusive stage – which it can do – the ultimate crushing charge, designed to wholly discredit us, is that we are emotional. Emotion is unmanageable, unclassifiable, unreasonable, untrustworthy. It is illogical. A terribly prominent and respectable elderly gentleman thus castigated me many years ago; flushed purple, sweating, trembling, he brought his elegantly manicured fist crashing to the table. "You naturalists are nothing but starry-eyed idealists – you're nothing but *emotional*!" he panted. That should have put me in my place for keeps.

I happen to loathe and abominate blood "sports." I think that killing any sensate being for recreation – for fun – is evil and contemptible. I have said so, for public consumption, many times. The most frequent theme in the resulting letters I receive is that I have absolutely no rational argument to present, and that as the result I (sneakily) resort to purely emotional appeals. Some of the mail, by the way, has to be opened with my asbestos gloves. I find this interesting.

In San Diego, California, there is an institution called "Sea World." This is a superb marine aquarium – probably the best I have ever seen – that is severely marred by a number of nauseatingly anthropomorphic captive dolphin, whale, and sea lion shows. (In these shows the animals' tricks and routines are without exception described as "behaviours" by the ringmasters. Either this is to bring a kind of clinical scientific aura to the performances or to intimate that the demeaning acts are "natural"; I am not sure.) In any event they are no whit different from the dressed-up chimp shows of my childhood or the hat-wearing cycling bears of the Moscow circus, which I find embarrassing and saddening. Yes, I *am* emotional. Aren't you?

At the San Diego "Sea World," last time I was there, there was prominently displayed the following message:

THE EXISTENCE OF ZOOS, OCEANARIUMS AND
AQUARIUMS IS BEING THREATENED!

You and your children may not be able to have the thrilling experience of visiting a zoo or oceanarium much longer.

Harmful and restrictive legislation is threatening the continued existence of these institutions, long dedicated to the protection and conservation of animals.

Much of this legislation – hastily conceived, often poorly interpreted and rigidly enforced – has been proposed by influential, emotionally oriented, pressure groups.

Through a lack of information and cooperation, their good intentions have resulted in often harming the very animals they sought to protect.

ZOOACT, formed by members of the zoological community, can provide a way to halt this threatening trend – before it's too late.

There is much in this to analyze, including among other things the claim that such institutions contribute to the protection and conservation of animals – a claim I have never seen substantiated. I have already made some reference to that. Relevant at this stage in our discussion is the careful and deliberate injection of that dirty word again. Our emotional orientation, it would appear, comes close to being a crime against nature.

What is this pejorative use of the word "emotion" really all about? Even the most casual investigation seems to show that emotions evoke fear. I think it is fear of the chaotic. To be emotional means to set aside organization (reason) and to drift perilously toward chaos — or at least to allow it to intrude sufficiently to threaten the edifice of reason. Like many others, the conservationist fears personal and wider disorder; like many others, he values his reasonable universe. Naturally, our opponents appeal to the public emotion in order to condemn our emotion, turning disadvantage into instant advantage. In so doing, they respond to and project their own worst fears — with human emotion.

Since emotions cannot be explained rationally, the wildlife preservationist is once again saddled with an impossible burden of proof. We don't even know what emotions *are* — merely that they exist. But that existence can only be shown within an experiential context — never by rational explication. There is no such thing as an emotion in isolation; it must have a context. You have to love or fear or hate *something* — even if it is only yourself. Unfortunately for all of us, the technocratic-scientific bowdlerization of "reason + experience = knowledge" into "reason + rationalization = knowledge" ensures that the experiential evidence, soft though it may well be, cannot be admitted.

Here is what I have referred to as our "double-bind." In the desperate attempt to show at any cost that we are not emotional, that we are entirely reasonable, we mount such incredibly twisted "rational" arguments as some of those reviewed earlier. We accept the onus of projecting and sharing that strange world view which is our own most intransigent difficulty. Wildlife preservation is for emotional, intuitive, irrational folk – women, for example. And so it goes. Every challenge we attempt to meet on its terms, rather than on our own grounds, creates new and even more recalcitrant problems. Our fear of appearing to be emotional requires us to fabricate our own rationalizations lest we invite the terrible censure of the technocratic age by committing its one unpardonable sin – irrationality.

If, as I suspect it to be, the root fear is of disorganization, then appropriate vigilance must be maintained for the general security. The conceptual edifice of rationality, in spite of itself, *knows* (reason + experience) that it is vulnerable, and acts accordingly. Thus we have orthodoxies and the threats of reprisal that protect them. If there is no split between man and nature, then man is no more than an animal – and you know what *that* means. It means, among other things, no human uniqueness in the matter of immortal souls. And that means total disorganization of the edifice, for keeps. Close ranks, brethren, your self-interest is on the line.

The only alternative to closing ranks is to open the kingdom to all comers. There is some precedent for this; women can vote now, in some places. In the time frame of human history, it was only an eyelash-blink ago that Europeans were justified in putting to the torch creatures that *looked* human but obviously were not because they had never heard of Jesus Christ. Even more recently it required a monstrous bloodbath to determine that one man has no right to own another. Eventually, however, even women, Indians, and blacks have been accepted into the human species and granted some but not all of the perquisites of membership.

It has been pointed out that the fearful atrocities of earlier times and of our own day are made possible in large measure by a conceptual-perceptual trick known as "pseudo-speciation." You don't napalm people; you napalm gooks. Thus we have wogs and niggers and spics and kykes and all the rest of it. Such are the uses of symbolism. The murderer on the loose or in the dock is an "animal"; social ideals and ethics no longer apply. Pseudo-speciation frees us from rules against killing, cruelty, and other uncivilized "animal" behaviour. Man/nature.

It may very well be that the rational fear of emotion is more intimately related to the human fear of nature than we like to think. Clearly the Rational Being occupies his place on the power pyramid because it is in the universal nature of things that reason is more important than non-reason and thus that man is more important than non-man. (This of course is a classic after-the-fact rationalization; it would not have been made in australopithecine times.) The fear of disorganization of the power structure is as profound as the fear of personal disorganization – perhaps even more so.

Fear of disorganization (breakdown) is fear of the loss of control. The applicability of this in the individual emotional sense needs no further emphasis. I would suggest, however, that the identical fear sustains the conceptual man/nature dichotomy. That which is not under control is threatening, whether it be the backyard crabgrass, the spruce budworm, the source of crude oil, the stock market, or oneself. An acquaintance of mine in the arctic town of Inuvik once said to me, "John, we've got to do something about all these ravens here in town!" "Why so?" I asked. "For heaven's sake, man, look around you – there are so *many* of them! They're going to be out of *control!*"

Loss of control is the abdication of power. It is tantamount to chaos. The universe is orderly, therefore chaos is unnatural.

The universe operates according to ultimate laws, therefore chaos is immoral. That which cannot be controlled (managed, manipulated) may be seen as both unnatural and immoral – psychopathic. The unmanageable is that "animal" side of us that surfaces in criminal court from time to time. The unmanageable is also the mountain peaks and the ocean abyss and the farthest ice-packs and the rain and the drought and all other such items as have not yet been boxed into the management chart of the human organization. So it is seen that the ravens of Inuvik (prospering on our garbage) are thumbing their amiable beaks at universal order and thus at us.

We resent and fear disorder, the absence of control, the breakdown of authority, the weakening of predictability. I believe that the individual fear of death is also the fear of disorganization. It is culturally conditioned, and it works. As we have seen, there is no individual immortality imperative, only a species one. Where our culture sees to it that the species immortality imperative is kept solidly in place, and it can do nothing about individual mortality, it does maintain a deep and abiding *fear* of mortality in the individual. Fear of the *experience* of death is not innate in us; it is created and sustained entirely culturally, which means rationally. Death is the final sting, the ultimate victory of uncontrollable, unmanageable, immoral, chaotic *nature* – from which experience we are snatched at the final exhalation by the gorgeous rationalization. Spirit over flesh, man over nature.

No example of the role of rationalization in sustaining the man/nature and reason/emotion polarities can compete with this one. It is the conclusive denial of the role of experience in knowledge. It is the conclusive act of closure. It is the final and absolute rejection of human membership in life process, the last unilateral decree of divorce from sentient Earth, the transcendent taxonomy. No longer do we have the jesuitical life kingdoms of plants, animals, man; now we are back to two, but a different two — biotic being, and man. The apogee in dualism is the perigee in self-delusion.

The belief system is quite unable to accommodate the experiential content of the classic knowledge equation. Such experiences as emotion and feeling are admitted into the model of reality only where to deny flatly their existence would be palpably irrational, as for example in dreaming or other common experience.

If, as I believe them to be, rationalizations are short-term "hypes" that are injected into the vacuum of empty experience, then all the wildlife conservation arguments in the world that are mounted from a would-be "rational" base are predestined for failure. If wildlife preservation really *is* for its own sake — which means for the experiencer's sake — then there can never be any "reason" for it. There is no rational argument for experiencing; it is above and beyond all logical capture.

The experiential vacuum is not part of our biology. It is entirely a matter of cultural conditioning, of nurture. There is no limit to experience in us, so far as I can detect, save what we cold-bloodedly and self-consciously impose upon ourselves in response to the conceptual tyranny of western "civilization." All is not lost, however. Once we recognize that there really *are* cultural (purely artificial) barriers to our experience, it is possible to begin to entertain the possibility of breaching them. Our minds *can* be opened to qualities of experience that will eventually allow us the state of being that is wildlife preservation. Only one act is required, and that is perhaps the most difficult (and yet, paradoxically, the easiest) you have ever undertaken. That is the act of opening.

Opening is a tricky business, and as in any other act of compliance, it gets more difficult if you allow yourself to brood about it. As the churchmen well know, you never can tell what might fly in. I think of the cultural belief system as a great white soft swollen belly. Incise it, and much will tumble out. The root dread, however, is of contamination that might *enter*. This dread pervades all of our institutions. It is why ecology has sometimes been called the "subversive science." Now, many a proselytizer will urge you to "open up"; he means, of course, just enough to allow the insertion of his particular offering, against which the hope is that you will have no antibodies.

To open is to make yourself consciously and deliberately vulnerable. It is an act of acceptance, an act of compliance. Opening may take a modicum of courage, but so did your very first public speech, your first dive off the high board, your first driving lesson, your first parachute jump, your first acceptance of mortality, your first serious doubt.

Opening is conscious and deliberate compliance with what *is*, not with what is imagined. It is the acceptance of whatever comes, the discarding of comparative "values," the denial of no possibilities, the entertaining of all possible impressions. Thus experience of wild nature becomes experience of our selves. It is no longer the perception (imaging, matching, rationalizing, identifying) of the world, but of *being* it. This is no mere empathizing; it is experiencing.

FIVE

The STATE *of* BEING

The other morning – just a day or two before this page went into my typewriter – I was awakened smartly at dawn by the close whooping morning song of a troop of gibbons. Rushing out the door, I very nearly fell over an unfamiliar blonde five-year-old in spectacles and blue pyjamas, sitting elbows on knees, chin cupped in hands, unmoving, intent. She glanced up somewhat sternly, as though to shush me, and softly whispered "Monkeys." I was rude enough to ask whether she could see the apes (I wanted to). She shook her head, wiping me from her world wholly and finally. At last I had the grace to sit and listen too.

The jungle shrieked and shook and crescendoed and echoed and re-echoed, and a mynah bird picked up the theme. I have heard many a dawn chorus in many a place, but I shall never forget that one, that quality. The overwhelming primate celebratory songs, the green dripping rain forest, the faint glowing early light, the intense little kid, the toads and the frogs and the newts, and – once again – free flow.

When I was very young the experience was entirely

accidental, unpredictable, ungovernable. It came and it went, and much of the anguish in the awareness of separation was caused by one's lack of control of free flow. You couldn't hold it, and you couldn't command it. It just happened – and then it was gone. A frequent fear was that it might be gone for good; but it never was. It is still here, still alive and well, and thanks to some non-rational, experiential, unreasonable, acultural, non-intellectual, emotional contrivance that defies all logic, in the right circumstances may now be summoned almost at will.

A day or two before the gibbons and blue pyjamas episode I found myself at 2.30 in the morning on a vast moonlit beach watching a seven-foot-long leatherback sea turtle dig a deep pit in the loose sand, lay forty-two eggs in it, cover the nest, and laboriously make her way back into the shimmering South China Sea. Since I was watching in the usual schizoid way of the naturalist, marvelling at it all whilst at the same time paying keen attention to each minute detail and considering the chances of hatching success, hatchling survival, turnover, and all the rest of it, it was prudent to hold the free flow for later.

That moment came when the great reptile finally heaved her immense bulk to meet the first wavelet at the shore. A dancing splash wiped the grainy sand from one enormous eye. A shuddering abyssal breath, another ponderous lurch, and her foreflippers were in the water. The slope is gradual here, and the turtle was still treading sand for more than twenty yards. But the heaves became easier, more rapid, more fluid, and at last she floated.

She remained on the surface for several moments, possibly to catch her breath. In those moments, as gentle ripples broke soundlessly over the gigantic carapace, now wetly silver in the moonlight, my universe was no longer mine alone. It was ours. With the leatherback turtle, I slipped beneath the smooth surface of the sea . . . As the ceaseless cycle turns, universe without end, life without beginning, pulse without destiny, form without function, love without self, circling among unknown stars, we are once again *in touch*.

I do not know how long I stood there in the shallows, but when I again perceived the ocean in a "waking" way the turtle was long gone. I was quite alone. The group of which I was part had moved a little distance down the beach to watch another leatherback going through the same ancient procedures. Though I too watched the second one, there was really no need to, and I knew it. All had been accomplished and incorporated – finally and forever – already.

Now, my point in reporting all of this is not to apply one more layer of mystery (mysticism) to the wildlife experience, but rather to emphasize that when I say that the fate of the sea turtle or the tiger or the gibbon is mine, I mean it. All that is in my universe is not merely mine; it is *me*. And I shall defend myself.

I shall defend myself not only against overt aggression, but also against gratuitous insult. There is a splendidly designed polar bear exhibit at the Metropolitan Toronto Zoo, on the outside wall of which is a plaque announcing that the exhibit was made possible by Imperial Oil Limited. I have mentioned already that my archetype of the ugly, offensive, and disgusting is the 45-gallon fuel drum. Since Imperial Oil Limited is one of the prime "developers" of the Canadian arctic, I interpret the tastelessness of this particular exercise in the context of the claims of zoos with respect to the protection and conservation of animals.

All, however, is not gibbons, leatherback turtles, and polar bears. For most of us, these are "exotics." They are important, but they are not everything. No individual species or group of species is everything. As I have tried to point out, the limits of "everything" turn out to be the limits of individual internalized experience. And the potential for this is limitless. Free flow is available to you anywhere – if you allow it.

Domestic animals are not part of my operational definition of "wildlife," but they are sensate beings, and many pets are willing to accept you into their states of being if you behave circumspectly and appropriately. This can be greatly helpful as a beginning. For a long time I and a number of others loved a tough old tomcat who with muscle and know-how ruled absolutely and for a surprising number of years several blocks in old residential Toronto. Also, surprising for a tomcat, he died not of wounds and injuries but of sheer age. He was aloof and self-contained, and he knew all there was to know, but he was tolerant of humans. Rarely, he would allow communication at the level of which I speak. One example will do. Very late one night, I was occupying one kitchen stool, he another. Everyone else was asleep. Suddenly, out of long, long silence, he stirred and *spoke* to me. What he said was brief and not susceptible to translation, but something prompted me to put my head to his, to open, to listen, and to accept. Something prompted me to *comply* – to my everlasting profit. For now the scarred and tattered and lumpy old puss and I can resume our free flowing circuitry whenever either of us wishes.

There is absolutely nothing unusual about this experience. Anyone who has ever loved a nonhuman being knows the extraordinarily encompassing sense of unity that is possible, at least occasionally. I am certain that you know the feeling as well as I do. I suggest, however, that rather than an end in itself this experience is merely the beginning – the opening of a gate that leads to the very nature of being, the nature of life experience and of life preservation. All I ask here is that you allow yourself to extend this selfless "identification" –

for that, essentially, is what it is — beyond those individual beings that you "know" in the conventional sense. Touch is good, but not essential. The animal need never know of your presence, for the bond is ancient and indestructible, and *you* are playing the active role in opening your world to his. So long as you are able to crack the plastic cultural capsule sufficiently to allow *his* universe to flow into yours, you have begun to grow. Put your head to the head of the tomcat; the tiger is near. The final step is accomplished when you can find (retrieve) the tiger in your universe at will, and the tomcat is no longer required as facilitator. Let him sleep; you know now what his dreams are made of.

Steal one moment away from your striving, and look at the cock pigeon strutting on the eavestrough. Look at him; *see* him. That wildly staring ruby eye looks a bit crazy, and it would be if it were in your head, but it's in that pigeon head; it belongs. Absorbed as he is in his display ritual, he may slightly lose his balance, careen for an instant in an undignified way, then regain his composure. It looks funny, but he doesn't fall. You would be killed. See the bird; really *see* him, and *feel* the urgency and the perfection and the beauty of his hot being. Forget your self-consciousness and your stereotypes; set aside your image-cards and accept what is right in front of you. Know that you and he pulse as one, and that you always did. You may not look the same, but you and the pigeon are no more than differently assembled manifestations of one thing, one process, one being. *Experience* that state of being, passively. Recall that state of being the next time you hear of someone poisoning pigeons. You too will be sick at your stomach.

In Vera Cruz you can buy a rock band consisting of stuffed bullfrogs, each holding a tiny instrument. In Nairobi you can buy an umbrella stand made from the foot of an elephant and a fly-whisk made from a zebra's tail. In Bangkok you can buy a stuffed mongoose and a stuffed cobra, locked in eternal embrace. In Puerto Vallarta you can buy two stuffed pond turtles dancing a fandango. In T'aipei you can buy chessmen and ear-rings carved from elephants' tusks. In Hong Kong you can buy an aphrodisiac made from powdered rhinoceros horn. In Kuala Lumpur you can buy a jungle butterfly embedded in plastic on a key chain. In Frobisher Bay you can buy a narwhal tusk. In Caracas you can buy a shrunken monkey head.

In the alchemist's dungeon that is almost any well-appointed shopping centre in the "developed" world, you can buy cosmetics, transmission fluid, and pet food made from whales; you can buy the hide of a lynx in the form of a hat, or gloves made from the skin of an unborn lamb; you can buy a coat made from seal whelps; you can buy a tropical finch in a metal cage and a Siamese fighting fish in a plastic bag; you can buy firearms and whammo ammunition, and multiple hooks with barbs on them; you can buy sharkskin shoes and the unspawned eggs of a sturgeon; you can buy the pulverized enlarged liver of a force-fed goose and the testicles of a bull and the brain of a calf; you can buy chemical biocides and plant growth inhibitors; you can buy the sterile eggs of an untrod chicken and the tongue of a feed-lot steer that spent its last weeks hock-deep in its own manure; you can buy medicines made from the blood and viscera of living laboratory animals.

You can also buy the Holy Bible and the Declaration of Human Rights.

As every wildlife preservationist knows to his cost, the simple minded among us are wont to interpret our anger as misanthropy. I try to tell them that we are honest folk; if we were misanthropists, we would say so. Still, it is often charged that we "care more for animals than for people" – again an unwitting emphasis of the "man/nature" problem. This, as I perceive it, is cultural constipation at its fullest. It is impossible for such as these to shift – even momentarily – outside the cultural hallucination sufficiently to see that what we care about is life – all life in all forms – or, perhaps better expressed, living. I care for the phenomenon of living more than I care for the contorted belief system of my culture. I care for wildlife preservation more than I care for its opposite – the technomachine and that which serves it. I care most deeply of all about the failure of wildlife preservation in my lifetime.

At the outset I hypothesized that those arguments we have mounted for wildlife preservation have been unequal to the task. Initially, my faith in that hypothesis was grounded simply enough in the widespread accelerating decline and disappearance of wildlife and wild places. At least on that evidence our arguments seemed insufficient. So they were put to the test; we looked at the most prevalent groups and families of arguments in order to try to see what it was about them that made them so apparently ineffective. This led us to certain underlying (root) problems that appeared to become even more important than the mere inadequacy of our stated case. Indeed they seemed to militate against *any* kind of argument. Argument might not be a factor after all.

In other words, the opening hypothesis seemed to have been correct, but for reasons other than my original ones. Most of my starting assumptions have been revealed for what they were – unfounded assumptions. We have seen that there can be no "rational" argument for wildlife preservation, just as there can be no logical explanation of quality experience. It now seems to me that argument *itself* – in the sense of reasonable dialogue – is not only inappropriate to our subject matter but also may be destructive of it. There is no "logic" in feeling, in experiencing, in states of being. Yet, these same phenomena appear to be prerequisite for wildlife preservation.

One of the more peculiar traits of our society is its assumption — its insistence — on solutions. Just as there are reasons for all things, so there are solutions for all things. Always there are ultimate answers; there is no problem that is not amenable to logical reduction. This, as we have seen earlier, in spite of such bewildering enterprises as ecology. I have no "solution" to the wildlife preservation problem. There may not be one. But given the somewhat shaky assumption that one exists, I sense that I can at least feel the direction.

The direction seems to lie in the compliant acceptance by individual human beings of membership — which is to say, "place" — in the beauty that is life process. This is too easy for me to say, for the difficulties are formidable. Such an act — a conscious and willing act of self-recovery — would require the recognition of options that have long been masked from us by our accumulated tradition. It would require the healing of perceived dichotomies, the mending of conceived alienations, the redirection of our fabricated imperatives toward reconciliation with that long-forgotten quality that is the nature of being. It would be the dissolution of the ancient western divorce from nature.

Throughout this essay I have called attention to the many symptoms of our malaise — our discontent, our uneasy rationalizations, our gnawing ambitions, our terrible paranoia — with respect to our perceived purpose in the scope of things. I interpret these and other symptoms as illustrating the frustration of a fundamental biological (and thus human) drive toward wholeness, toward belonging. Implacably and relentlessly, our culture keeps us blinded to the qualitative sensory options that are ever at hand and infinite in number.

The fallacy that is conventional wildlife conservation is a flat denial of those options. I cannot know whether your or my qualitative experience of nature will ever benefit wildlife, but I do know that the joy of self-recovery and liberation from cultural tyranny is contagious.

One Cosmic Instant

Instant

A NATURAL HISTORY *of*

HUMAN ARROGANCE

"When the stars threw down their spears,
And water'd heaven with their tears,
Did He smile His work to see?"

— WILLIAM BLAKE

CONTENTS

AUTHOR'S FOREWORD

Frustration, dismay, and varying degrees of anger have been the inseparable companions of the wildlife conservationist always. It is a fine line between these emotions and misanthropy. Rather than opt for troglodytic retirement, however, I have chosen the route toward what I consider a reasonably civilized form of sabotage.

In the course of writing this, I have discovered that buildings and mechanical contrivances must be a great deal more susceptible to explosion than are concreted ideas. Most particularly resistant is the edifice which is collective Western homocentricity, bulwarked and reinforced as it is by apathy, greed, ignorance – and tradition.

Although I am very much aware that there is nothing one can know for certain, I have proceeded from one radical assumption. Contemporary Western man in the overwhelming majority considers himself fundamentally different and distinct from the living world that gives him both substance and sustenance. This imagined separation between man and "nature" has provided the conceptual framework for a further doctrine, that of absolute human power and authority over the nonhuman. These ludicrous but terrifying notions have become solidified in our collective thought in a ridiculously brief period of human and Earth history.

The conceptual power structure is recalcitrant. Its successful penetration and demolition will be essential to the continuance of life processes as they presently exist. I have attempted to illuminate in perspective a few of the more obvious building-blocks between which the fuses must be insinuated.

~ **O N E** ~

The **TEMPLE** *of the* **MAYA**

Massive cream-and-sable king vultures have been wheeling above the Temple of the Giant Jaguar for centuries. Red, blue, and orange-carunculated heads cocked pale eyes downward at the great gleaming plaza of Tikal, where a brilliant civilization shone momentarily, only to be reclaimed by the oppressive lowland jungle of Guatemala.

The Maya knew the vulture, and carved his likeness on stone. Oc, the vulture, knew the Maya, and he watched the Maya vanish. The vulture still circles overhead. Parrots screech and toucans croak from the canopy of towering mahoganies, and ocellated turkeys gingerly make their way across openings on the forest floor. Except for the absence of the Maya and the emptiness of his shrine, little has changed.

In a cosmic sense, nothing has changed at Tikal, for civilization is ephemeral. Silently the king vulture drifts across a gap in the forest cover around the ruins. Once, a culture as sophisticated as the world had ever seen flourished here. During Europe's Dark Ages, the Maya knew about astronomy and mathematics, and had a concept of time.

But cultures are fleeting. They evolve — sometimes with extraordinary suddenness — and wither just as quickly. On the other hand, the king vulture was the same animal before, during, and after the brief scintillating moment of the Maya.

I — a birdwatcher — left Guatemalan Mayaland with far more than a list of new tropical species. I left with a sense, not so much of the vulnerability of human cultures, but of their susceptibility to unexpected, rapid change. In thinking of the relative immutability of the king vulture as a species over a time-span of mere centuries, I was struck by the vast difference between the slowness of natural selection and the speed of the rise and fall of cultures. I came to brood on human aspirations, and on their potential for taking new directions.

If the naturalist values the objects of his obsession, he is compelled to involve himself in the world of people who at this moment are watching, with indifference, as the king vulture and his tribe gradually disappear. Like that of the Maya, something in our society is grievously wrong.

The nonhuman world is important to the birdwatcher. Its relative importance grows, in inverse relationship with an inevitable misanthropy. One's disillusionment with society's treatment of nonhuman nature is built on a body of evidence which is conspicuous on every hand.

One of the commonest questions I am asked is: "What 'turned you on' to birds?" When I reply, truthfully, that I cannot remember, I am rarely believed. People look for instant, simple causes for observed effects. It was too long ago for me to be able to recall; certainly it was before I was ten years old. Perhaps it is more enlightening to say that I cannot remember when I was *not* interested in birds.

Children respond to animals, and as a youngster I realized that a ravine, then unspoiled, near my house was full of wild animals, especially during spring and fall bird migrations. At those times I would conceal myself under a shrub, remain as inconspicuous as possible, and wait for the birds to come to me. But I had great trouble attaching names to them; they moved too quickly for me to pick up their distinguishing features. In my school library I had already discovered the stunning array of Fuertes and Brooks plates in Forbush's *Birds of Massachusetts*; I had discovered Frank Chapman, and Reed's little guides, but all of these had stern limitations for field identification. The living world of birds was explosively revealed to me in 1934, when I was eleven years old. Like Keats opening Chapman's Homer, I first looked that year into Roger Peterson's earliest *Field Guide to the Birds*. My response was not so much like that of someone watching a new planet swim into his ken as like one liberated into a new galaxy, equipped with the tools with which to comprehend it.

I especially remember one May morning, during a warbler "wave." That day there was an unusual number and variety of warblers. With Peterson's book in hand, I hid motionless in my retreat and watched the parade. I was surrounded by birds — darting, fidgeting, flashing, shimmering — a magnolia warbler, a black-throated blue, a black-throated green, a myrtle! And I could now *identify* the birds, which brought me infinitely closer to them.

From the ability to attach names to the hundreds of kinds of birds, plants, butterflies and other forms of life flows the realization that these new phenomena have about them a degree of predictability. One doesn't look for the brilliant larvae of the monarch butterfly outside the milkweed patch, nor does one tune an ear for the song of a robin until late in March. One looks in beech trees for the nesting excavations of pileated woodpeckers, and discovers that a tiny nest with an inner lining of horse-hair is probably that of a chipping sparrow.

It requires little ecological insight to learn that one must go to marshes to find bitterns, rails, swamp sparrows and marsh wrens, and to open fields and meadows for bobolinks and savannah sparrows. It becomes apparent that there are specific habitat requirements for birds. The moment some favourite location for a species is changed or eliminated by human activity, there are issues at stake. The boy who saw Cory's least bittern (a rare and local colour phase) driven from the marshes of Ashbridge's Bay in Toronto by a sorely needed sewage treatment plant learned early of the complexities and conflicts of conservation.

All is not orchard orioles and apple blossoms, nor purple finches at the backyard feeder. Though we are aware of human over-population and the attendant problems, not until their effects are seen first-hand are we struck by the immediacy and reality of the modern environmental crisis. One rages inwardly at the effects of sheer human numbers on the landscapes and the birds one wishes to preserve, and laments the degradation of the human condition. Awareness of what is happening to the things one holds precious leads the observer to question the human "values" which have brought the world of nature to its present level of subjugation. The things I value – such as birds – are being destroyed by other things I also value: human life. The conflicts and the paradoxes are staggering.

Not all in the birdwatcher's life is grimness and dismay. There is the lighter side. As his travels become more extensive and his virtuosity develops, he becomes able to practice ornithological "one-upmanship." While observing a grey-headed gull in the humid estuary of Ecuador's Guyas River, he might murmur, "Ah. Last time I saw that species was on

Lake George in Uganda, under the Mountains of the Moon." Or, "In the forest of Bialowieza in eastern Poland you can see all four species of European flycatchers in one day." Or, "Isn't it odd – the discontinuity in the distribution of the azure-winged magpie, which I have seen in both Portugal and Japan. Except for Mongolia and northern China, it does not occur anywhere in between."

More important than the birdwatcher's skill and knowledge is his awareness of the cyclic and repeating nature of life. The seasons are recurring and predictable, but things are never *exactly* the same from year to year. As I write this, it is mid-March. Outside my window a supercharged cock house sparrow has begun his courtship ritual, and is trembling with intensity. There is a small flock of slate-coloured juncos on the dirty snow near the back fence, foraging beneath reddening osier canes. A cardinal is singing lustily a couple of gardens away, and the backyard skating rink is melting fast. Two or three crows have gone by in the last few minutes, the advance guard of regiments to come. There is a great flutter of copulating pigeons on a chimney-top across the way. The day is overcast, but the prospect is good. Any day I will hear the grinding scrape of a grackle's song, and my faith in the continuity of things will be reaffirmed.

This spring, predictable though it may be, will not be like any I have seen before. No two springs are ever alike; things change in nature, and flux does not end with the circling seasons. One cannot anticipate everything. Still, I am saddened by the certain knowledge that this spring there will be fewer mallards and canvasbacks than there were in other years. There will be fewer blue whales – almost none, in fact. There will not be as many leopards or pelicans as there were in the springs of my youth. I will probably not see a peregrine falcon this year. Because of real estate development on their wintering grounds in the Bahamas, there will be fewer Kirtland's warblers in Michigan than there were a year ago. But there will be seventy million more people on Earth than there were at this time last year.

One is drawn into conservation in any number of ways. A farmer becomes aware of soil erosion and soil depletion and water shortages; an angler becomes aware of water pollution and eutrophication, hydro dams, and over-fishing; a bird-watcher notices the steady disappearance of the habitats that sustain his objects of interest, and the poisoning of those natural habitats that do remain. The common factor in all of these environmental changes is man.

Something in the biosphere is drastically out of synchronization with everything else. That *something* is man. One need not be a prophet of doom, nor even a mildly conventional pessimist, to see that there are components in the biospheric life system which are dangerously out of hand, and that time to restore the bond with nature is limited.

I have spent a lifetime in conservation affairs and have gradually but inescapably been convinced that up until the present time all we conservationists have been able to achieve is fight a delaying action. The best we have been able to accomplish has been the temporary treatment of symptoms. We have not reached the root cause of environmental distress, which is Western industrial man.

Western civilized society appears to be indifferent toward nonhuman nature. We have seen the bison, the trumpeter swan, and the bighorn sheep fall before the gunners; we have seen the prairie dog, the black-footed ferret and the whooping crane give way before the sod-busters; we have seen the giant baleen whales reduced to the vanishing point by international commercial greed. Most significant of all, perhaps, has been the unchanging traditional assumption that although the loss of these animals may well have been regrettable, it was inevitable and unavoidable in the context of the advancement of human progress. It was sad that these things had to happen but, after all, what was at stake was human welfare, and in the final analysis human welfare comes first. That's only "natural."

The World's Fair at Montreal, Expo '67, was subtitled *Man and His World*. To this day its amusement-park successor remains *Terre des Hommes*. In 1967, my thoughts on man and his world drifted back to Nicolaus Copernicus.

Nicolaus Copernicus was born about 500 years ago, and in the course of his life he demonstrated the theory that the sun – not man – is at the centre of our planetary system, and that the Earth and the other planets revolve around it. His system was later developed by the likes of Kepler, Galileo, and Newton. However, Copernicus had merely resurrected and improved upon a notion that had occurred to Pythagoras of Samos, who was born more than 2,500 years ago. Yet, the theories of Copernicus and Pythagoras have had little impact on modern man, who still terms it, "Man and His World."

This reflects an ancient and profound view with regard to the role of the nonhuman component of the biosphere. It is the root cause of the despoliation and disfigurement of nonhuman nature which has now become the human environmental crisis. The conservation movement is doomed to failure unless Western man summons the courage and the will to challenge his fundamental assumptions.

Nowadays, public attention given to environmental questions generally concerns pollution – the contamination or overloading of air, water, soil, and living tissue by both natural and synthetic substances in volume or of kind with which natural systems cannot cope. This is a symptom of a much more deep-seated illness. Pollution *can* be dealt with.

The *causes* of pollution are far more difficult to approach. Pollution and its effects constitute a rash of conspicuous abscesses which are the external manifestations of something mortally wrong within.

It is of little consequence to the conservationist whether one's initial interest in nonhuman nature is for pleasure or the advancement of science. What is relevant is that, if for no reason other than his own survival, man must soon adopt an *ethic* toward the environment. "The environment" encompasses all nonhuman elements in the one and only home we have on Earth. However, it will be some time before we are able to enunciate, much less promulgate, an environmental ethic because, fundamentally, the ethic runs contrary to our cultural tradition.

Ethics have been associated with man-to-man or man-to-society relationships. They have not been concerned with man's relationships to the nonhuman. Most moral philosophers have not acknowledged that man might have at least some ethical responsibility to the nonhuman. Perhaps this is because we cannot conceive of having any ethical responsibility to that which is not capable of reciprocating. Ethics, morals, fitness and propriety of behaviour – these are human attributes. They do not exist, so far as we have been able to determine, in the nonhuman world. (That this may be a mere problem in communication does not seem to have occurred to us.) Since ethics do not exist in the nonhuman world, there is no need to apply them to that world. Our attitude toward the nonhuman world is not immoral: it is amoral.

In this book I concentrate on Western European cultures, not because those cultures have had any "corner" on ethics (they most conspicuously have not), but because Eastern ethics are becoming less and less relevant in the face of world industrialization in the Western image. Japan is probably the best current example of the abandonment of an ancient and nature-oriented ethos for the spirit of progress, Western-style.

The achievement of an ethic to guide our behaviour toward the nonhuman world will inescapably involve the brutal violation of our cherished beliefs. It will necessitate close examination of our attitudes and assumptions, and of their origins.

Man's imagined removal from nature had its beginnings with the very dawn of culture. Whatever culture is, it is not biological. It is an artificial, conceptual inheritance, transmitted over the generations in the form of traditions, beliefs, revealed truths. Cultures change as they are passed on by the generations. They evolve at a much more rapid pace than do biological organisms. Therein lies our hope, and Earth's.

We cannot change our biological inheritance, but we can and do change our cultures — consciously. Conscious change of direction toward the environmental ethic will mean the practice of a kind of artificial selection — choosing certain positive elements in our traditions, and rejecting negative elements. The selective process will not be easy, for it will demand something that is foreign to us — humility. It will demand unprecedented humility in the face of the encountered facts of the biosphere and the cosmos. It will demand willingness to see ourselves in the perspective of time of infinite duration and of events of unimaginable magnitude.

TWO

ONE *in a* MILLION

The realization that I was "one in a million" burst on my consciousness years ago on encountering the astonishing fact that one human ejaculation may contain a quarter of a billion male sex cells. On later reflection, considering the potential number of ejaculations in a father's lifetime, the number of female sex cells produced, and the infinite possibilities of their combinations, I became keenly aware of "unique self." Even now my mind still boggles at the enormity of the odds against this peculiar individuality, and on the extraordinary nature of chance. Speculating on the outcome of even one sexual union, much less thousands, is as futile as attempting to count the stars in the night sky, to say nothing of the planets.

People tried to count stars long before it occurred to them to try counting sex cells. Both are fruitless; there are simply too many of them. Easier to reckon with are such phenomena as comets, which are often conspicuous and spectacular, and some of which are predictable. The most famous of these, Halley's Comet, has a cycle of 76 years, and if you miss it once, you are not likely to have a second chance.

Halley's Comet last appeared in 1910. An old friend of mine, then at university, planned elaborately for a special viewing. On a particularly auspicious evening, he collected his girl friend, together with a blanket to sit on, and some refreshments, and repaired to the top of a nearby hill. It need not be added that, on that occasion at least, the comet was not seen. In his latter years, my friend (who by that time had become a distinguished ornithologist) would grin somewhat ruefully and shake his head, in full knowledge that he would not be around in 1986. He was fond of saying that some opportunities only arise once. And he was perfectly right, whether one's time-scale is human or cosmic.

The expression "Spaceship Earth" is in wide use nowadays, chiefly as a means of emphasizing the finite nature of our planet's life-support systems. On viewing Earth from a distance, as moon-voyagers have done, its isolation in the vastness of space is awesome. Neil Armstrong has remarked that he had not been involved in environmental issues until he observed Earth from afar, and recognized how relatively small – and how alone – it is.

Most astronomers estimate that our personal space vehicle has existed in its present form for about 4,500 million years. In other, embryonic and developing forms it has been around for considerably longer than that, but some element of controversy has surrounded its prenatal career.

Considering the mortality of species, faunas, planets, and solar systems, and the ephemeral nature of constellations and galaxies, the question of whether the universe has always been there or not is insignificant; it has little relevance to the transitory condition we on Earth call "real life." Real life on Earth will long be gone before there is any substantial change in what we now see in the heavens, and no one will be here to witness the outcome.

Our Earth spaceship is one of 100,000 million stars in one of the 1,000 million galaxies that lie within the range of our restricted observation. In the context of the one new solar system, the odds were weighted heavily against the emergence of life on one of a handful of planets, but in view of the number of suns and their satellites in our own galaxy – much less the universe – perhaps the appearance of living beings no longer seems remarkable.

Like all familiar living organisms, solar systems and planets have their beginnings, middles, and inevitably their ends. They are conceived, they develop, and they become extinct, both as individuals and as groups. It had to start somewhere, however, and it would appear that it began with the primal presence in the space-void of the most abundant element that exists – hydrogen.

In the immense black spaces between the planets and between the stars, there is little of anything, save a loose scattering of hydrogen atoms. They are not sufficient to form even a puff of vapour, for at this instant in cosmic time most of the supply of hydrogen is locked up in the sun and in the substance of the planets. Long before the birth of the sun, however, there seems to have been in this corner of our galaxy a certain concentration of hydrogen atoms which was sufficient to form a cloud of gas.

Here, the force of gravity came into play. Where the gas was sufficiently dense, gravity drew the component particles of the gas closer and closer together. The gas cloud began to form a ball. As mutual attraction between the particles increased, the ball became smaller, tighter, denser, until it shrank to less than one-millionth of its original diameter. The faster the particles of space-dust moved toward their centre of attraction, the hotter they became. Eventually the centre of the sphere became so hot (about 20 million degrees F.) that the atoms of hydrogen began to fuse. The original cloud of stellar gas had become a gigantic hydrogen bomb.

Though a relatively minor hydrogen bomb by cosmic standards, it is potent enough to generate the energy to sustain life on Earth. The thermonuclear process will continue until the sun runs out of hydrogen to fuse. As the supply of available hydrogen begins to dwindle, the sun will gradually begin to swell – at first imperceptibly, then more and more quickly. The planets will be consumed in orderly procession, beginning with the one nearest to the sun, Mercury. (Or, if there is a planet between Mercury and the sun, it will be the first.) Mercury will be followed by Venus, Venus by Earth, and so on. Scientists predict that after its swollen stage the sun will gradually shrink again, lose its heat and brightness, and disappear into the void from which it came. Observers in other galaxies will probably not notice its final twinkle.

The sun is now in the full warmth of its summer maturity – and so it must be to support life on any of its planets. The sun first acquired planets of its own because not all the hydrogen "dust" in local interstellar space went to feed the sun's thermonuclear plant. At a very early stage, the newly emerged sun was surrounded by a gigantic disc of galactic gas. The disc was far enough away to escape being drawn into the central inferno, but not so far as to escape the gravitational attraction of the sun. Gradually, these gases began to coalesce, in much the same way as the proto-sun had done, and to form the planets. Each gathered towards its nucleus whatever matter chanced to be drifting about, until there was little of anything left in what was now interplanetary space. Nine satellites had formed, and – *quite by chance* – one of them found itself at a critical distance from the sun.

The planet thus favoured was the blue one. The Water planet. There happened to be atoms of oxygen in the original gaseous inventory, in sufficient number to combine with the ubiquitous hydrogen atoms.

Had the new planet been ever so slightly closer to the sun, its water would have remained in the form of vapour, due to the proximity of the sun's heat. Had it been ever so slightly farther away from the sun's warmth, its water would have remained in the form of ice crystals. No accident of fate was ever more significant to Earth than its critical distance from the sun, allowing for the thread-thin range of 100° C. within which the existence of liquid water is possible.

At the formative time of its "condensation," Earth was extremely cold, and its water was frozen. Robert Jastrow has described Earth at this stage of its gestation as "a Neapolitan sherbet of frozen water, ammonia and methane, plus various kinds of rocky substances, all immersed in a gaseous cloud of hydrogen and helium." However, a combination of factors which may have had to do with its own radio-activity, or with constant pounding by space-particles, resulted in Earth gradually warming. It warmed so much, in fact, that it melted. The ice crystals liquefied, boiled, and vaporized. Minerals, now molten, sorted themselves out according to their relative specific gravities, although the eventual distribution of what would be land masses was a long way in the future.

In the due passage of time, Earth began to cool off, starting naturally at its surface. The outermost layers of material congealed and solidified, and the cooling process slowly worked its way inward, leaving the molten core Earth retains today. But it was hot enough at the surface that the water supply was still in the form of vapour, which must have totally enveloped the evolving planet.

Since Earth's inception, the process had taken about 50 million years. Earth was a hot, but cooling, ball, surrounded by dense layers of clouds. But when the temperature finally dropped below the critical 100° C., the water vapour began to condense, and the rains started. How it must have rained! When we consider the enormous quantity of water now on Earth, the deluge from the skies must have been torrential. Presumably some of the rocks at Earth's surface still retained enough heat to vaporize much of the water, but eventually condensation overtook vaporization; and the result was our water cycle.

The lowering temperature caused Earth's surface to contract and to wrinkle. Unevenness at the surface resulted in the development of low catch-basins which soon filled with water running from the higher ridges in cataracts. Where it collected over wide areas, the embryo oceans developed. The sun, warming the surface of the seas, caused water to evaporate and to be carried into the primitive atmosphere, to be moved by weather systems around the globe until lowered temperatures caused it to condense and return to Earth as rain. Rain entered the rivers and thus the seas, and the hydrologic loop was complete.

Still, there was no life on Earth. The first atmosphere did not contain molecular oxygen – the product of green plants of the land and phytoplankton of the sea. Before these "producing" organisms, the atmosphere must have consisted of a mixture of water vapour and various other primal gases. And, as though waiting to be put to use, were immense quantities of liquid water at the surface of the planet.

The oldest recorded living things are organisms resembling bacteria, which have been found in Precambrian rocks in South Africa. These are fossils believed to be more than 3,000 million years old. The oldest fossils that are recognizable as plants are those of some algae found in Ontario; they are about 2,000 million years old. Other alga fossils, which may be somewhat older, have been found in the Rocky Mountains. These ancient plants are the earliest known organisms which were capable of photosynthesis, and they were probably the first sources of molecular oxygen and thus the originators of the secondary (breathable) atmosphere. Pouring the breath of life into the air, they made possible the emergence of the biosphere.

Where did these "pioneer" algae come from? Where did their progenitors originate? The basic "building blocks" of life are proteins, made of combinations of about twenty different kinds of amino acids. Scientists have synthesized some of these amino acids by firing electrical charges through such gases as ammonia, methane, water vapour, and hydrogen, all of which are thought to have been contained in the original, primary atmosphere. They were also in the original cold cloud from which Earth condensed. A spark of lightning may have begun the long "chain reaction." Ultraviolet radiation from the sun could have been a catalyst. By these or other means, or a combination of them, the ultimate experiment got under way.

Precipitation is our only source of fresh water. The original molecules of water were formed, it is thought, by the combination of hydrogen and oxygen atoms in the first cold cloud. The process may well have continued ever since, and the global water supply may be perpetually renewable from space. Some water molecules drift toward Earth from outside, but since most of the hydrogen atoms are components of the molecules of matter in general, and there are few of them loose in the atmosphere, it seems more likely that the water supply is finite, and whatever form it may be in at the moment – solid, liquid, or gas – the water we now have is all we ever had and may expect to have.

From oceans, lakes, streams and ponds, and from plant leaves and people, evaporation lifts about 100,000 cubic miles of water from Earth's surface for recirculation every year. Such is the force of solar power in the biosphere.

Although the water cycle is by far the most impressive of Earth's life-support systems, there are others which are equally vital to the maintenance of living organisms. These are the carbon and nitrogen cycles, which make their unique contributions by way of the soil. Good soil is filled with multitudes of micro-organisms – among them, bacteria and fungi. These "plants" do not possess any of that miracle-stuff, chlorophyll. So they must operate on chemical energy instead of solar energy. These soil organisms are the power sources for the carbon and nitrogen cycles.

Green plants extract carbon dioxide from the ambient air for the manufacture of carbohydrates, releasing oxygen in the process. The carbon compounds thus retained in the structure of a plant get back into the soil when the plant withers and falls. Or the plant may be nipped off, before its time, by some grazing animal. In this case some of the carbon which was in the plant may be exhaled by the animal as carbon dioxide, and some of it may become part of the animal itself. In the latter eventuality, the carbon must await the demise of the animal before it can return to the soil. When that happens, the micro-organisms in the soil decompose the structure of the animal. The carbons enter into the bodies of these minute animals and plants, with an eventual return to the soil. Some of the carbons are oxidized, and thus get back into the air. The result is that the carbon component of living things is forever being cycled through the soil and into the air for subsequent use by plants.

Plants must also have nitrogen, but they cannot extract it from the air. So the nitrogen cycle consists of its being made available to plant roots in solution with other material, by way of the soil, again due to the contribution of micro-organisms in decomposition. An animal eats the plant, the animal dies, and the cycle is continued.

About three-quarters of the breathable atmosphere consists of nitrogen; the rest is mostly oxygen. The extraordinary thing about that part of the atmosphere which can support plants and animals is its thinness. Earth is about 8,000 miles in diameter and 25,000 miles in circumference, but there are only about seven and one-half miles of gaseous exchange above its surface. When puffing at the summit of Mount Everest (29,141 feet), or enjoying aperitifs in a jetliner at about that altitude – less than six miles – the air pressure is below one-third of that at sea level, and one cannot breathe without artificially supplied auxiliary oxygen.

Since the oxygen in the atmosphere is a biological product – the exhalations of green plants and of phytoplankton – everything ultimately depends on what goes on at the surface of the globe. The soil cover is, of course, much thinner than the breathable part of the atmosphere. The continents are said to be covered with soil to an average depth of from only seven to ten inches. Yet that micro-thin envelope has sustained itself and myriad living creatures for billions of years.

Earth, a relatively minor planet in terms of size, is by our standards extremely large, yet the biosphere is so thin as to be virtually transparent. It has been pointed out that if Earth were reduced to the size of a billiard ball, its mountainous surface would appear ivory smooth. We just might be able to recognize the great continents, but the oceans would be "mere films of dampness." We would not be able to detect the biosphere at all.

Slight and delicate though it is, the biosphere is sufficient for every life purpose. Miraculous? Perhaps not. Unique? Almost certainly not. In human terms, however, it is all there ever has been and all we can hope for. Earth – the product of sheer cosmic chance – may or may not be one in a million, but for our purposes it might just as well be.

The CONCEPT of COMMUNITY

In the lexicon of a group of scholars, the word "community" has a meaning vastly different from that given to it by a builder-developer – or by a sociologist, for that matter. Or by a global tribalist. To a biologist, community is something still different.

"Community," in the context of natural history, is a concept, not an absolute. The definition of any natural community is arbitrary and abstract for it depends always upon the scope in which one views it. For one man, observing the activity in a terrarium, that is a perfect community. For another, the community may be a pond; for another, a woodlot. For still another, the community may be the broad stretch of the Sargasso Sea.

The simplest community to *comprehend* is a lichen. There are many kinds of lichens throughout the world, most of them growing in such bleak and inhospitable surroundings that they may be the only sign of life. Lichens manage to eke out a living in some of the most biologically impoverished parts of the world chiefly because a lichen is not *a* plant, but a community. And a community is more than the sum of its parts.

Lichens are formed by a symbiotic relationship between two different kinds of plants, fungi and algae. The fungus contributes something and the alga contributes something; together they mutually benefit in what must be the world's most basic community. Lichens look vastly different, being of all colours and shapes, but each is a partnership in the very best sense. The fungus provides the framework or the foundation for the little cooperative enterprise. It secretes acids which help the lichen to maintain a foothold, and gradually draws mineral nutrients from beneath it. The alga, in turn, feeds the community. Algae have chlorophyll, the vital substance for the process of photosynthesis; fungi do not. Together, the two plants thrive as one plant; neither could exist without the other.

This interplay or interdependence between two forms of life in the same place at the same time constitutes a biotic community. It is simple in its number of components, but remains a real form of ecological cooperation.

Cooperation manifests itself on every scale throughout the natural world. One of the most remarkable cooperative communities that has ever evolved is one in which the aggregate constitutes what would appear to be one individual animal – the Portuguese man-of-war. This handsome "creature" is familiar to almost everyone who has beachcombed. Occasionally one will find dozens of small blue "balloons" scattered along the beach, where a group of the tiny sailors has been stranded by an onshore breeze. Although the blue (and sometimes pinkish-trimmed) bladder is the most conspicuous part of the man-of-war, attached to it there are slender, filament-like tentacles, which may be as much as fifty feet long. These tentacles are poisonous, and can deliver a fearsome sting. The man-of-war feeds on small marine organisms, which are knocked out or killed outright by the stinging tentacles, and are then digested in the "platform" which forms the lower part of the animal's float.

Like the individual lichen, the Portuguese man-of-war is not *an* organism, but a great many organisms living together in a cooperative community which supports all of them. There are at least four different kinds of animals involved, none of which could exist without the others. One animal is the float, or sail. Presented to the wind, it moves the man-of-war about the ocean. Another animal is the digestive apparatus. Still another looks after reproductive matters. A fourth is represented in the stinging and feeding tentacles, each one of which is an individual being. Far from being one creature, the Portuguese man-of-war is a community of animals, none of which could survive independently.

Lichens and men-of-war are composed of only a few kinds of participants. It is a short step, however, to communities of much greater complexity. A small terrarium may contain hundreds or even thousands (if we consider soil microfauna) of organisms, all dependent upon each other, all contributing to the success of the "undertaking" as a whole. The woodlot is another step up the scale of complexity, but the principle remains the same.

There are few free-loaders in natural communities. Some may give the impression of enjoying an unpaying ride at the expense of the societies with which they coexist; there are parasites on plants as well as on (and in) animals, and although they may live at the cost of their host, in the final act of its individual drama each of them contributes its remains to the nourishment of the community.

Some communities may appear more complex than would seem necessary. In a grassland community, for example, we expect to find primary "specialists" who carry out basic functions. There will be tillers of the soil, whether they be nematodes, earthworms, pocket gophers, ground squirrels, aardvarks, or elephants. There will be primary converters of plant energy, vegetarian grazers such as horses, cattle, antelopes, kangaroos. But in some favoured parts of the world which have not yet succumbed to human pressures and where the natural way still prevails, we encounter more kinds of animals doing these specific jobs than we might have expected.

On an East African grassland there may be as many as a dozen different kinds of hoofed animals all living in a common area and all appearing to be involved in the same activity. Why this overlap of primary converters? On close observation, the realization emerges that although these different vegetarians *appear* to be doing the same thing in the same place, they are in fact not. The zebras are feeding in somewhat longer grass than most of the antelopes are. Wildebeest (gnus) are grazing on areas of somewhat shorter grass; it has already been "mowed" to the proper length (for gnus) by the zebras. Some of the little gazelles are grazing where the grass is even shorter, after the gnus have passed by. They are cropping the very smallest shoots. Each of them, specializing in its own way, is occupying a particular food niche, exploiting the resources of the environment in a slightly different way. The community has a place for everything – so long as everything stays in its place.

The system becomes misplaced only when two or more kinds of animals in a given community attempt to earn their living in precisely the same way. The structure of the community allows no room for such competition. Garrett Hardin puts it succinctly: *"Competitors cannot coexist."*

In nature, competition has none of the connotations of cut-and-thrust between individuals which human societies

have permitted to evolve. Natural competition is an impersonal struggle for existence between species, without any
consciousness of the struggle.

Unnatural or artificially induced competition is quite
another matter. In parts of East Africa, chiefly that area generally known as Masailand, one can clearly see what happens
when there is unnatural competition for a food niche in the
grassland community. Domestic cattle have been widely introduced and are maintained in unfortunately large numbers.
Their impact on the vegetation has been calamitous. Grazing
at ground level, they concentrate on one food source – grass –
building up unnatural pressure on the ground cover, which is
soon eliminated. Erosion follows quickly. The impact of wild
animals, on the other hand, which evolved not only with each
other but also with the entire grassland community, is spread
over the land both horizontally and vertically. Each specialist –
from the toy-poodle-sized dik-dik to the eighteen-foot giraffe
– is engaged in its own endeavour, and the pressure is never
too great at any point in the food conversion process. Left to its
own devices, the community functions smoothly. When a
foreign ingredient is introduced, an unnatural form of competition results in gradual but certain breakdown.

There is more to a community than its biological component. There is the extra dimension represented by what Odum
calls the "nonliving component" – the gases of the air, the
minerals of the soil, and of course the water which sustains it.
When we add this second major (nonbiological) component,
what began as a community becomes an ecosystem.

Like a community, the size of an ecosystem is arbitrary.
Some ecosystems are easy to define, as they are in the case of
communities – a marsh, a forest, an island. However, it is
impossible to describe an ecosystem in spatial terms; like the
community, it is a concept rather than an absolute.

We live in an age of "systems." In essence, there are only two kinds of systems: open and closed. Almost all systems are open; they depend for their functioning on inputs from the outside. The terrarium or aquarium, considered as an eco-system, is open; at least occasionally it needs air, water, or both, from the outside. The pond is an open system in that it depends on both water and air from outside. A spacecraft is an open system. In the absence of a recycling technology which would allow its occupants to live indefinitely on their own wastes (as inhabitants of the natural world have always done), the spacecraft must take essential supplies with it. The only closed ecosystem — functioning completely on its own — is the biosphere. It can be argued that the biosphere is not closed either: it depends upon the constant and uninterrupted supply of solar energy from the centre of our planetary system. In a biological sense, however, Earth's systems are self-sustaining.

What each living thing depends upon is the constant flow of energy. Energy must pass through three fundamental stages or trophic levels. One consists of the green plants, which receive energy from the sun and fix it in the form of food-stuff for themselves and for all other living things. (This level also includes aquatic phytoplankton.) The second level consists of all animal life. No animal can fix solar energy; it must eat either plants or animals which have themselves eaten plants. Thus plants are producers of energy and animals are consumers of energy. The third trophic level consists of organisms called "reducers," the microscopic life in the soil — the fungi, bacteria, and so on — which break down the structures of dead plants and animals, releasing their components to the cycles of the biosphere.

The general trend of natural communities and ecosystems is toward complexity. In the case of animals, we usually regard those which are more complex, or more "highly" organized, as the most sophisticated and successful products of evolution. In the same way, we see the more diverse and complexly interconnected ecosystem as the more stable. The more parts it has, the better its chances of remaining on an even keel for a longer period of time. A community such as a Portuguese man-of-war or a lichen can find itself obliterated should even one of its members be lost. A more complicated community, such as a woodlot, can, through shifting checks and balances, cope with almost any *natural* upset of one of its components.

Homeostasis, the state of equilibrium within a community, is achieved and maintained through constantly adjusting mechanisms within the structure which tend to compensate for fluctuations in any part, such as overproduction or overconsumption. There are constant shifts in emphasis and changes in numbers of plant and animal species. For example, cyclic increases and decreases in numbers of snowshoe hares are followed by corresponding fluctuations in lynx numbers. But the trend is always toward stability, with subtle adjustments constantly acting as though to keep the air bubble in the middle of a spirit level.

Since natural communities and ecosystems are continually moving toward multiplicity and diversity in order to keep the homeostatic level steady, the simplification of ecosystems is contradictory to every ground rule.

The simplification of landscapes is especially conspicuous in modern farming methods. In place of the scores of species of grasses which once were the base of the luxuriant prairie community, there is now a monoculture of wheat – one species of grass – which extends in every direction for hundreds of miles. Not only is this monoculture uninhabitable by most wildlife species, but it is also extremely vulnerable to attack by plant-eating insects and by plant diseases. Uninterrupted stands of grain – or of any other plant species – are open invitations to blights of every kind. Such bonanzas do not occur in nature.

Nowhere is there a better example of the simplification syndrome than in Canadian forestry operations. In the eyes of the forest industry, one of the most fearsome agents of destruction is the spruce budworm (which concentrates primarily on balsam fir). Its larvae eat the spring growth buds of the trees, and they have devastated great stretches of commercial forest. One of the reasons for the rapid spread of budworm infestations in recent times has been past cutting practices, and historic (one might almost say primeval) forest management techniques. Like the modern big-scale industrial farmer, the forest operator is interested only in certain species. To him, the best forest is the one that consists not only of a few commercially desirable species, but of stands of roughly the same age, so that cutting rotation may be carried out with greatest speed. The result has been enormous stands of balsam fir which, in much the same way as with wheat, encourage infestations of "pests."

In nature, trees of all ages would be intermixed with trees of many different species, thus lessening the danger of pestilence on the grand scale. To exacerbate the forestry problem, after a widespread budworm infestation there is the added danger of fire, due to the many dead trees. Even natural fire caused by lightning can thus be made more severe by human agency. Forest practice invited the budworm which prepared the way – the tinder – for the fire.

The simplification of natural systems, whether they be prairies or woodlands, is the inescapable result of the human drive toward specialization, in current times aided and abetted by economic forces. If he is in the business of growing wheat, the farmer grows all the wheat he possibly can, at the same time eliminating those plants he does not want, and which he calls "weeds." In so doing, he may well be removing the very buffer against depredation that his wheat crops need.

The commercial forester, interested in one tree or another, or a handful of species, attempts to eliminate "weed trees" in order to make his operations more economic — more competitive. The old-line traditional forester has been trained to examine the parts, not the whole. The holistic approach to forest ecology is unknown to him. Since "specialist thinking" pervades almost every aspect of human affairs, it comes as no surprise that comprehension of the totality of life has been slow to dawn on us.

Admittedly, one cannot "see" an ecosystem. No one stands back and says admiringly, "Now *there's* a lovely ecosystem!" But that should not mean we cannot comprehend it, or grasp the concept of community.

In the context of human affairs, we bandy the idea of community, but we are unable to see the human species in the larger scheme of the biological community. Man fails to acknowledge that he may be dependent on things other than human ingenuity.

In pondering the "community," we must allow our thoughts to range more widely than the human community — into the filigree of interrelated and interacting elements which form the system in which we live.

One may view a natural ecosystem in terms of a modern corporate phenomenon, the "conglomerate." A wide and diverse base of activities helps to ensure the stability of the enterprise as a whole. If the bottom falls out of ssts or petrochemicals, there are always hotel chains or distilleries. If one strand of the web weakens or snaps, there is always enough resilience in the structure to compensate for that loss, enough to swing into a new configuration of equilibrium, with the stresses distributed so evenly over all the other strands that no undue strain comes to bear at any point. Provided always that there is a sufficient variety of components.

The longer an ecosystem has existed, the greater the time that has been available for diversification and complexity, and thus for stability. Time is the essential ingredient for the development of stable communities. Obviously, the first rank in world ecosystems is occupied by the oceans. They have existed longer than any other communities, and they contain an incalculable volume and variety of living things. In addition to its plants and animals, the sea holds minerals and other materials delivered to it from the land masses in quantities we can only guess at. Its energy flows from phytoplankton (microscopic marine plants) to zooplankton (equally minute marine animals) through a bewildering array of lesser and greater invertebrates, fishes and mammals, through whom the cycle of life passes back to bottom-feeding decomposers on its return to the "great chain of being."

One of the ocean's most important contributions to the biosphere is oxygen. It has been estimated that as they go about their photosynthesizing activities, the phytoplankton release up to 70 per cent of the world's oxygen supply — besides being the fundamental base of the marine food pyramid. Recently it has been learned that even relatively small quantities of DDT, released into the sea from world river systems in concentrations of only a very few parts per billion, can reduce the photosynthetic capabilities of some kinds of marine phytoplankton. The conclusions are chilling. The oceans are vast, and in their antiquity (and thus their diversification and stability) they have been able to accommodate most of the insults which man has lavished upon them. That no longer appears to be the case. Byron declaimed that man's control stops at the shore, and that on the watery main all deeds were those of the ocean itself. The shadow of man's ravage is sickeningly evident today in the greatest of all eco-systems, from the sluggish shining oil slicks on its surface to the invisible but deadly synthetic molecules in its bottom mud.

The forests (especially in the tropics) offer the next widest variety of living beings in any given area. The typical equatorial forest consists of layers – from the great green canopy of the largest trees through the shade-tolerant forms which stand beneath them to lesser trees, shrubs, and smaller plants of the cool and luxuriant floor. Each stratum has its characteristic birds, insects, reptiles – even mammals. Such a forest is a welter and delight of sound and movement. One can almost hear the staccato heartbeats of those who live at such high-key intensity.

Forests are never static. Even with no human inter-vention, a forest is constantly in a state of transformation. Trees grow to their fullest extent, mature, and die. When at last they fall, the sunlight streaming through the gaps thus left in the canopy promotes the immediate growth of other species which theretofore had been inhibited by the shade. Later, fallen logs give both sustenance and substance to yet other species. New kinds of trees, new groups or associations of species are constantly "striving" for "succession."

Forest succession has been found to be reasonably predictable. After fire, flood, or clearing, it is possible to anticipate the kinds of small plants, shrubs, and eventually trees that will appear. They may not be the same species that were there before. The nature of the soil, exposure to wind and sunlight, and other factors all combine to determine who will succeed whom in the procession. In theory, the forest eventually reaches its "climax" – a steady-state beyond which it cannot go. The theory of climax must of course be viewed in perspective. Given a stationary state, with the influencing factors remaining constant, a forest will reach an "optimum" condition, dominated by trees best suited to the circumstances which predominate.

But things in nature do not remain constant, and the concept of climax must not be taken as an absolute. If North America is still in the midst of a warming period following the last (Wisconsin) glaciation, then the boreal forest of spruce, fir, and larch will, because of changes in climate, eventually move northward. To the south, it will be succeeded by the mixed and broadleaf forest in an orderly procession northward. On the western plains, it will be followed northward by the prairie grassland. The evergreen forest will "invade" the arctic tundra. *If* it is warming. If, on the other hand, we are about to enter a new glacial period, and the world climate should cool, then the reverse process would take place. These changes are of course imperceptible in the context of the human lifetime, but should we be present long enough to observe them, we would see a conspicuous process of succession over periods of thousands of years. "Climax," then, is useful in describing the dominant vegetation on a given site at a given point in time; but in reality – over the millennia – it is no more than an abstraction. For if the pure steady-state existed in nature, there could be no evolution.

On seeing an immense stand of forest, one often concludes that the soil must be rich and nutritious in order to produce such spectacular crops of greenstuff. In the tropics, unfortunately, this presupposition has led to many disastrous errors in land management. Generally speaking, tropical soils are not all that productive. In East Africa, the unglaciated soils have long since been leached of most of their nutrients by the annual rains. Where there are forests, the trees are shallow-rooted in the thin soil, and they survive because in the humid warmth the decomposition rate is high, and there is a fast turnover of nutrients. But when trees are cut, and the thin soil is exposed to sun, wind, and water, erosion occurs with shocking rapidity. Flying over mountains in South America and Africa, one is appalled at the devastation that follows removal of the forest cover. In many parts of the world, the substitution of agriculture for forest can only result in ecological and thus human hardship. Conditions prevail in natural situations for sound and established reasons, and we tinker with nature at our own peril.

Because they have existed for so long a time, oceans and forests are varied and stable ecosystems. At the other end of the scale are places that are surprisingly young – so young that they have only begun to realize their potential ecological diversity. Perhaps the best example of these, with the exception of volcanic islands in the ocean, is the arctic. Most of the arctic today is released from the grip of Ice Age glaciers for only a few short weeks each summer. In contrast to the forest, where innumerable generations of trees and shrubs have contributed their nutrient-filled remains to the food cycle, the arctic has virtually no soil at all. There has not yet been time for the weathering processes of nature to break down and distribute the rocky rubble left by the glaciers. Sometimes only inches beneath is the unrelenting permafrost, the permanent layer of year-round frozen earth, gravel and rock (or ice) into which water, and the roots of plants, can scarcely penetrate.

One major result of the permafrost and consequent poor drainage is rapid run-off of the spring and early summer meltwater. The formation of new soil is delayed for protracted lengths of time. These factors, together with the relative paucity of sunlight over the year, have limited the plant growth to tough sedges and grasses, and, of course, lichens. In addition, such "trees" as willows and alders scatter over the tundra north of tree-line proper, forsaking height for sturdy stems and ground-hugging form.

The tundra does, however, support a surprising number of animals. There are few species, but in some cases great numbers of individuals. Hordes of biting insects appear in summer, as far north as there is anything warm-blooded to feed upon. There are ground squirrels to dig wherever there is sufficient soil to dig in, and there are grizzlies to dig out the ground squirrels. Soggy tundras and deltas attract some of the world's most notable congregations of ducks, geese, and swans. And there are the unbelievable herds of caribou, which from the air remind one of the antelope which move annually over Tanzania's Serengeti Plain. No two parts of the world could be more different than the arctic tundra and the African veldt, yet unrelated species of animals behave in startlingly similar ways, in remarkably parallel adaptations to their respective habitats.

The solar energy budget of the arctic is so constrained that it requires vast stretches of landscape to sustain only a few animals. The herds of caribou (like the African wildebeest) cannot afford to be sedentary; if they were, they would over-graze and eliminate their own food supply. They keep moving, and their impact on the "reindeer" lichens which sustain them is spread thinly and evenly over vast stretches of landscape.

Arctic tundra ecosystems are critically vulnerable to human disturbance. The Eskimos who lived in the arctic before the invasion by Europeans were hunters. Their impact on the homeostasis of the system, like that of most predators, was minimal. Today, however, with the advent of industrial man in the north, things may be expected to change at a rate heretofore unknown in nature. One of the gravest dangers to the arctic – paradoxical though it may seem – involves removal of the permafrost. Although its presence inhibits the development of luxuriant plant cover, the removal of the permafrost has devastating results.

When tracked vehicles move over the tundra in spring and summer, they sink into the softly thawed insulation of the ground cover, leaving trails. These tracks, open to twenty-four hours of daily sun, allow warmth to penetrate the denuded vegetative cover to thaw the permafrost. As the thawing proceeds, the tracks become wider and wider, runoff exposes still more permafrost to the sun, and the process continues to a point at which the track of one vehicle begins to resemble a muddy 1917 Flanders trench, though considerably longer. The eventual ecological implication of all this, which is occurring on a massive scale, is presently under investigation. One conspicuous result of rapid meltwater runoff thus induced is the simplification of the complex network of tundra tarns and ponds, which may be expected to have implications for nesting waterfowl.

Damage to the ground cover is becoming depressingly conspicuous in parts of the arctic now under exploration for fossil fuels. Aesthetically, the vehicle scars are deeply offensive, but it is difficult to put a price on unsightliness. Ecologists do not yet know the full extent of the damage to the tundra ecosystem, which is such an excruciatingly thin envelope over the all-pervasive permafrost. Perhaps before the arctic terrain is scarred beyond recognition by industrial invasion, and before its marine community also succumbs, there will be a sufficient body of scientific knowledge available to permit intelligent planning of northern exploration.

The arctic ecosystem is as yet relatively simple. This is largely due to its youth. As time passes, a warming trend may bring new plants and animals to enrich the environment, to provide it with additional checks and balances, and thus with increased resistance to upset. A cooling trend may lock it all away, far from the reach of industrial man.

Through our failure to acknowledge our oneness with the total of other populations of biological organisms in our ecosystems, we reveal the egocentricity which has brought the world environment to the point of crisis. Reluctance to face facts which may be ego-destructive is self-deception. We have gone so far in our self-deception that the separation of man from his ecosystems has become even more than a deception. It has become an article of faith.

~ᴈ FOUR ᴈ~

The NUMBERS GAME

Whether there is "purpose" in life is questionable, but there are definite trends. Biologically, the trend is toward diversity and variety, in order that the stable equilibrium of eco-systems may be maintained. Trends must involve a force, and the fundamental force in living organisms is reproduction. It would seem that the object of all living beings is to produce as many offspring as possible in order that a sufficient number may survive the exigencies of existence to replace those continuously being lost to natural attrition. In some parts of the world, men want to have many sons to ensure that there will be someone to care for them in their later years. In nature, the attempt is to deal with the overwhelming odds against the survival of any individual or group of offspring.

A population consists of all individuals of one species in a given community. In studying natural populations, there are a number of elements involved. How *dense* is the population: how many individuals of that kind of organism are there in the area being observed? In the case of some invertebrates, one might be counting numbers per square foot; in the case of redwinged blackbirds, numbers per acre; in the case of deer, numbers per square mile.

The density of a population of animals is partly a reflection of the species' birth rate. Animals and plants have spectacularly different birth rates, according to their ways of life and their roles in the community. Plants, which are the primary food producers and are eventually devoured by everyone in the community (including each other), must perforce have a high rate of reproduction. Vegetarian animals in general have higher birth rates than meat-eating animals, because as primary converters they too are suppliers to the food pyramid. Their "turnover" is high. To understand what is happening to animal populations we must know not only the birth rate of a species but also its rate of turnover, and this involves its death rate.

Assessing populations is difficult. Many organisms are anything but sedentary. They come and go. Populations in local areas do not remain static. Nothing in nature does. There are individuals who, for various reasons, move away into new communities, and there are those who wander into an area from outside.

Despite fluctuations, natural populations (with such exceptional phenomena as lemming cycles) give the *appearance* of being relatively stable. Birdwatchers notice minor or even major changes in some species over periods of several years, but they also learn to expect reasonably dependable relative populations of the most familiar species. It was once thought that some mysterious mechanism was built into a species which prevented its numbers from "exploding"; somehow the birth and death rate were counterbalanced. However, this is not the case. Changes in the environment of a species may have extraordinary consequences; one change might wipe out an animal completely, while another might allow it to increase at an incredible rate. This is precisely what has happened with the human species, and it has been observed in many others.

There is a valid analogy between the uncontrolled growth of human populations and the unnatural increase of nonhuman species which man has introduced into new environments. Man has manipulated his own habitat to the point of creating a new kind of environment expressly suited to his wishes, and this new and special set of conditions has involved the elimination of most natural population controls. The same phenomenon occurs when we introduce a bird or other animal into a set of environmental conditions which is foreign to it. In most cases, the new species fails to survive because it is not adapted to the exotic situation. Occasionally, a new species may find itself able to cope with foreign conditions, and free from the usual biological controls in its original habitat. In such cases, the result can be impressive.

The deliberate introduction of the European starling and the house sparrow to North America, and of the rabbit to Australia, shows the consequences to animal populations in the absence of homeostatic controls. Virtually overnight, the birds and the rabbits increased to "pest" proportions in their new freedom from natural restraints. That is because the birth rate of most animals is as high as local conditions will allow. Among many small invertebrates and fishes the odds against the survival of an individual egg are of the order of thousands to one. In the spores of fungi, the numbers may be astronomical.

In one breeding season, most birds "attempt" to raise as many young as the traffic will bear – or as many as is physiologically possible for them. The major limitation on bird increase is the availability of food for their young. The constraints exerted by food supply cannot be demonstrated better than by the hawks and owls. Most birds do not begin to incubate their eggs until the entire clutch has been laid, and thus all the young hatch at the same time. But with birds of prey, all eggs are incubated from the moment they are laid. Since eggs are produced on successive days, or at longer intervals, the brood will consist of young hatchlings of varying ages and sizes. In times of plenty, all will survive. In meagre times, only the strongest nestlings (usually the first-hatched) will make it. When food shortage becomes acute, only the largest chick will come through, often by devouring its nestmates. This is a form of homeostatic population/food adjustment which operates according to the environmental pressures (or the opportunities) of the moment.

Smaller birds produce all the young of which they are capable and "hope for the best." The most familiar example among the songbirds is the backyard American robin. In the latitude where I live, robins generally nest twice in the course of a spring and summer, with a normal complement of four eggs to each clutch. If all of these hatched, and if all the young were to survive, we would have eight first-year birds and two old birds back in the garden next spring. Ten where there had been two. Games of arithmetic do not prove much, for exponential curves exist only on paper, not in nature. In theory, however, if robin reproductive success were to continue at this rate without interruption, with adults and young surviving for ten years (not an unreasonable age for a lucky robin), there would be 19,000,000 robins in the garden at the end of that fateful decade. And there would be an equal number in the garden of my neighbour. But this of course does not happen. Next spring there will again be two robins in my garden, not ten. Like it or not, God *must* see the songbird fall. In fact, He has to see to it that something of the order of 80 per cent of the robins are dealt with in a normal year.

There are abnormal years and abnormal series of years. In the middle and late 1950s, when the impact of DDT and other chlorinated hydrocarbon insecticides was beginning to be widely noticed, there was a notable die-off of robins. DDT sprayed on elm trees, in attempts to control elm bark beetles and thus the imported Dutch elm disease, was carried to the ground on leaves in the autumn. There it was transferred to earthworms which ate the leaves, thence to robins. George J. Wallace, studying the robins on the campus of Michigan State University, estimated that ten such earthworms, loaded with DDT, were sufficient to kill a robin. He reported that in only four years his local robin population on campus dropped from 370 to three. During the same period, robin populations in general were seen to dwindle over substantial areas.

The robins eventually picked up again, and robin populations seem to be relatively stable once more. But in the interval – after the bottom had dropped out and the long climb back had begun – somewhat more than 20 per cent of the robin population was allowed by the environment to survive in a one-year period.

Under natural and undisturbed conditions, there are a number of factors which tend to limit outbreaks of the proportions we have experienced with house sparrows and starlings, and which we could in theory experience with robins. One of these is crowding. Some large birds, such as eagles, resist the company of their kind (for reasons having to do with foraging space), to the extent of being entirely solitary in their nesting habits. Other birds seem to "demand" the most intimate togetherness. Great colonies of auks, such as murres, and of penguins and other seabirds are common. Some bird species will not come into breeding condition, or will not breed, without the stimulation of a dense surrounding colony of their kind.

A point occurs beyond which these seabird colonies do not appear to grow – as though an invisible dome were placed over the colony, allowing no further expansion. Certain gannetries in the world are known to have increased substantially over a period of years, but this is probably the result of protection. With the pressure off, the colonies have reverted to their "normal" size, where they will level off again.

In colonies of albatrosses, gannets, or auks such as puffins, the "core" of the colony usually consists of mature birds which have bred successfully in prior years. Upon arriving at the colony at the outset of the breeding season, these older and more experienced birds take up the choice nesting sites in the centre of the area. There, they are less vulnerable to predation, and, perhaps, more stimulated by the influences of sociality. The younger, inexperienced birds take up positions around the periphery of the colony, where they are more exposed to accident, and where they may not realize the benefits of a pleasant degree of crowding.

These younger birds may or may not be paired. They may go through the motions of display and preliminary site-choosing, and they may not breed at all. (In some albatrosses, the birds are not sexually mature for six or seven years, and even then they may not nest successfully the first year or two.) The young birds go through this ritual "practice," or they loaf about the outskirts of the colony, not even going through the most perfunctory nesting ritual. They become observers, and they may observe more than we surmise. In some kinds of birds, however, breeding is possible even when one or both members of a pair are still in sub-adult plumage. In a disaster year, when for one reason or another a large proportion of the fully adult birds may be removed from the colony, these younger individuals may be recruited into the main body of breeders. In such cases, they may nest successfully, as though to quickly bring the population density back to the proper level. They represent a pool from which the species may draw in times of necessity.

The pooling of potential breeders is nowhere more evident than in colonies of fur seals and sea lions. Here, a relatively few fully adult males hold all of the females in harems which are more or less large. The younger or older or incapacitated bulls form bachelor herds of their own. At the first sign of weakness or vulnerability on the part of one of the dominant beachmasters, there is always another bull ready to do him battle and to drive him away and take his place. This, however, has nothing to do with population *density* in the strict sense, for all the females in the colony are impregnated, regardless of which bull may have done the impregnating.

Without evidence that fluctuations in birth rate contribute to the stability of bird populations, control must be exercised by the death rate. Death rates change conspicuously. These are probably highest when bird densities are at their peak (when the auk rookery is "full"), and at their minimum when densities are lowest. The most prominent causes of increased death rates when animal populations are high are disease, predation, and food constraints. Various diseases are known among mammals. On one visit to the Galapagos Islands I noticed that the local sea lions were more numerous than they had been six years before. I noticed also that the beaches were littered with dead sea lions. A mysterious "blister disease" was sweeping through the colonies, killing old and young alike. On the basis of unusually high sea lion numbers, one may speculate that the disease is a response to population density. One may also anticipate a new equilibrium in due course.

More familiar mammal diseases — especially those of the wild dogs such as wolves, coyotes and foxes — are distemper and rabies. It does not appear certain that these are devices of population control, for numbers of predatory animals are stringently controlled by the relative numbers of the animals they eat. The diseases are probably always present in population "reservoirs"; and in times of malnutrition or other adverse conditions, they may tend to eliminate a certain number of young animals and those approaching senility whose defences are lower. There are of course sporadic wide-scale outbreaks of both rabies and distemper, in which larger than usual numbers of animals may be afflicted, but these do not *seem* to be related to population regulation. Among all animals, disease generally comes into play as a controlling agent when there is a substantial "surplus" of a particular species. This surplus, because of lack of food or other factors, is more prone to illness.

Predation appears to be in much the same category as disease. Under stable conditions, predation generally strikes the young and the aged – the weak and the infirm. It is rare for any predator to significantly reduce its own food supply. The survival of predators may be larger than average in years of unusual food abundance and it may drop when the pinch is on. There is a clear relationship between the number of predators and the number of animals available to them for food. The controlling agent in the predator-prey relationship is the prey species rather than the predator. Again, this is a homeostatic control. As V.C. Wynne-Edwards remarks with regard to senile mortality, it "occurs at the age that suits the species concerned, more or less independently of the particular agency that happens to discharge the death sentence."

Discharge of the death sentence *must* be carried out. Among animal species there is no avoidance of population limitation. A provocative agent of population control is Hans Selye's "general adaptation syndrome," relating to stress. This works chiefly on adults in a confined area.

Desmond Morris, whose books include *The Naked Ape* and *The Human Zoo*, is an authority on captive animals. No zookeeper in his right mind, Morris feels, would dream of confining animals in such restricted quarters as human city-dwellers find themselves. Overcrowding in animals results in increased aggressiveness toward others of the same species, in quarrelling, and sometimes in death. This mode of behaviour will keep population increase in line. But it may go beyond that, to cut down on absolute numbers.

Stress need not build to the point of blood-letting. As Selye has shown, many changes may be occurring within the individual in a physiological way which may not at first manifest themselves externally. These changes may show up in the adrenal glands, the thymus, spleen, and sundry other lymphatic tissues, the red cells in the blood, the mammary glands, the reproductive system, and the fat reserves.

The most interesting homeostatic control of natural populations is the phenomenon of territoriality. Many kinds of animals show a tendency to appropriate a certain physical area of their environment and to defend it against others of the same species. They may do this as individuals, as breeding pairs, or as groups or bands. The practice is more conspicuous among birds than mammals, although it is also characteristic of certain fishes and reptiles.

Territoriality appears to be a response to fluctuating food supply. Wynne-Edwards believes that one of the effects of territoriality is the spacing out of individuals of a population of animals, and that the practice not only produces dispersal of the animals but also functions to control population size. John Hurrell Crook states that "this hypothesis is part of a broader theory which holds that social organization is essentially a mechanism for providing 'conventional competition' whereby numbers are regulated by homeostatic dispersal in relation to food resources." Conventional competition avoids physical combat which would otherwise erupt under the stress of overcrowding.

It was H. Eliot Howard, a British student of birds, who in 1920 revealed to the world the widespread prevalence of territories among small songbirds. Since that time, a growing number of people have been led to study territoriality, but knowledge of it – in its full complexity – is far from complete. There is little outward consistency about territoriality. In East Africa, spotted hyenas and Cape hunting dogs get their living in essentially the same way – by pursuing and pulling down such prey as antelopes. The hyenas are territorial, while the dogs are not. One must probe deeply to ferret out just what it is in the life style of an animal that determines whether that species will or will not be territorial. Sometimes the same species may behave differently in different parts of its range.

There are vast differences in the size of animal territories. They may consist of many square miles, as in the case of a pride of lions, or they may cover only a few square inches, as in the case of a murre or other seabird virtually rubbing shoulders with its neighbour. The lions' hunting territory must be large enough to maintain a stock of game, or at least to allow game to pass through. The murres' fish-hunting area is the open sea. The only terrestrial real estate a murre needs is adequate space within which to incubate its single egg.

Some animals maintain territories only during the breeding season, some the year round. Others hold different territories in summer and winter. Most of our familiar migratory songbirds in temperate zones, such as thrushes and warblers, are vigorous in the defence of their nesting territories during the summer, but during the off-season they congregate in close-knit flocks, with no sign of the territorial jealousy they exhibited only a few weeks before.

Whooping cranes, on the other hand, have breeding territories in northern Canada during the summer, and off-season territories on the Gulf coast of Texas in the winter. European robins, which are not migratory, have both summer and winter territories.

The defence of territory is often confused with aggressiveness. Such defence is merely the response of the individual animal to the intrusion of a stranger of the same species. As someone once said, ironically, "The bear is dangerous: when attacked, it defends itself."

Defence of territory takes many forms. Most small birds define the limits of their plots by means of song; they have favourite singing perches which are situated about the territory and which distinctly mark its boundaries. Many mammals, such as dogs and cats, leave scent markings to indicate their proprietorship. In addition, as though to reinforce their scent markings, wolves will howl and lions will roar. Gibbons whoop noisily, and howler monkeys roar and growl. In the case of wolves, howling serves in most instances to keep packs from ever meeting up with each other. Thus, the population of a given species spreads smoothly through favourable habitat. Territoriality probably keeps the populations of a species forever exploring new areas, thus pressing toward the ultimate limits of suitable environment.

Man has been emancipated from most of the controls which curtail the increase of populations in other species of animals. We are not bothered any longer by predators — at least most of us are not. Western man has few food problems, and relatively few from contagious diseases. We do not practice infanticide, nor are we cannibals. Most of us have even survived the effects of stress. We have escaped most of the constraints to which other animal species are inextricably bound. The result has been an unnatural increase in human numbers, to the point where we endanger not only ourselves but the systems which give us life. By the sheer weight of our numbers, we threaten that magnificent kaleidoscope of other living things which also depend on the natural systems.

One of the most instructive accounts of the growth of human numbers is Goran Ohlin's background paper for the 1965 United Nations World Population Conference, *Historical Outline of World Population Growth*. It has served as the basis for most of the following estimates.

During his earlier days as a Pleistocene hunter and gatherer, *Homo sapiens* was a reasonably well-integrated member of the natural community. At the beginning, he was probably little more efficient than most social predators (which is to say, no more than middling-efficient), and he must have had only minor impact on the numbers of the animals he hunted. But his hunting improved, as the large mammals of the late Pleistocene learned to their cost, and he began to superimpose a technological tradition upon his earlier culture. Human populations at this time may have been as few as one million, or as great — at the outside — as ten million. We shall never know precisely how many men there were, but we can be sure that they represented a drop in the bucket relative to the total biomass of the great Age of Mammals. It was not until somewhat later that man began to have an obvious effect on the landscape. The destruction of the natural world began with civilization.

If we date civilization from the "agricultural revolution," that brings us to approximately 10,000 years ago. The partial domestication of certain animals presumably came before and may have led to the domestication of plants. Even more important to the environment was the end of the age of nomadism. It was the end of man's natural integration with the land and with the biotic community.

Because they kept moving, wandering pastoralists must have had only slightly more impact on the land than did the bison. (Goats and sheep, however, can stay in one place until they eat it down to the bare earth, and no doubt this happened quite early. But cattle are not as deadly because they are less efficient at reducing grass to desert, and periodically they must be moved to greener pastures.) But as soon as man stopped wandering and began to tend crops, permanent communities were formed, and this was the beginning of widespread environmental degradation in the Mediterranean basin. From an absolute maximum of perhaps 10 million during Pleistocene prehistory, the numbers of people jumped to at least 50 million by classical times. Communities had also become groups of communities, super-communities had become city-states, which in turn had become nations and empires.

By the time of Christ, the world population had grown to something between 200 and 300 million. Bubonic plague and other epidemics reduced the population; but by 1650, when Europeans were busy with global exploration and conquest, the number had reached about 500 million.

In another hundred years, while Jean-Jacques Rousseau was yearning for a return to the Arcadia of our forebears, the world human population had reached 750 million. It is no wonder that a romantic misanthropy was pervading Europe at the time. By 1850, due largely to advances in sanitation, public health, and medicine, we had reached our first billion. Headlong expansion in the New World had followed the ages of discovery and colonization, and the industrial revolution; and in another hundred years, by 1950, the total had reached two and one-half billion. We have done rather well since, due to such agents as DDT. Today we stand at approximately 3.7 billion — a rise of over one billion in what used to be considered the standard human generation. By 1980 it could easily be 6 billion.

The absolute numbers of people are striking enough, but the *trends* are even more impressive. Before the agricultural revolution, it is believed that the human population could double only in from 10,000 to 100,000 years. In the period between the agricultural revolution and the time of Christ the population doubled every 1,000 years. Between the year one and 1650 there were the effects of the Black Death, and the population doubled only at the rate of once every 2,000 years. But it picked up quickly, and doubled again in only 200 years between 1650 and 1850. It doubled again in the next 100 years. At present, the estimated doubling time is between 35 and 40 years. Yet there remain in our midst an astonishing number of people who would apply standards and criteria to human reproductive behaviour which may not have been appropriate even in the Dark Ages.

One of the greatest ironies is our humanitarianism. I have held in my arms a Guatemalan Indian child dying of malnutrition, and such is the nature of our species that I would have done anything in my power had I been able to save that baby. Many are being saved. Control of communicable disease, improvement of sanitation, control of infant mortality, massive advances in geriatrics, have all resulted in "death control." Pasteur, Jenner, Lister, Fleming, Banting, have helped alleviate the general human condition. Our population almost doubled in the short interval between Paul Müller's award of the Nobel Prize for his discovery of the insecticidal properties of DDT in 1939, and the 1962 publication of Rachel Carson's *Silent Spring*.

It is impossible to be indifferent to the story of DDT. It *is* possible to be ambivalent about it. I am distressed by the effect of DDT on the bald eagle, the osprey, the peregrine falcon, and the brown pelican; but I was also distressed when a malarial mosquito in Africa's Rift Valley assaulted me. Had I been weak or undernourished, the mosquito could have killed me, or so debilitated me that some other disease – perhaps the flu – could have finished the job. No doubt the mosquito *should* have killed me, because there are too many of my species. There is no doubt that that is the malarial mosquito's function. Whatever the consequences, DDT has saved innumerable people from insect-borne diseases. This has resulted in many more people.

Vasectomy – male sterilization – is suggested as the best contraceptive device. It is not very encouraging, however, on the basis of a little mental arithmetic. Generally, the operation takes about half an hour. An eight-hour day, with an hour off for lunch, represents fourteen vasectomies per surgeon per day. A team of three surgeons, working around the clock on eight-hour shifts, could perform 42 operations per day. Let us imagine further a clinic with ten operating tables and thirty surgeons, and our count comes to 420 sterilizations per day. Assuming no holidays or days off for this dedicated team, we come up with 153,300 operations per annum. How many clinics of thirty surgeons would be needed to take care of even one-tenth of the world's estimated 1,850,000,000 males? To say nothing of an equal number of males projected for the next 35 years.

Man has already exceeded the "carrying capacity" of planet Earth for his species, and in so doing he has crushed the life from species that, in the Pleistocene, were his brothers. There has been a breakdown in homeostasis. Despite the brash confidence of those who promote agricultural technology and the "Green Revolution," and despite the best intentions of those who call for the redistribution of wealth and food, nothing is working.

In his historic address to the United Nations in 1965, Pope Paul said (in translation), "Your task is to ensure that bread is sufficiently abundant on the table of humanity, and not to favor artificial control of births, which would be irrational, in order to diminish the number of guests at the banquet of life." Life in underprivileged countries is no banquet, His Holiness notwithstanding. Also, no banquet was ever crashed by a babe in arms: it is seared into my memory that the dying Guatemalan infant did not ask to be there.

FIVE

BALD EAGLES *and*
POT-BELLIED PIGS

If, as would seem to be the case, modern man has existed in recognizable form for about a million years, it means that that particular evolutionary experiment has been going on for about one three-thousandth of the history of life on Earth, and for about one four-thousand-five-hundredth of the history of Earth itself.

Richard Carrington has used the device of compressing time scales to help our perspective. Let us imagine the entire history of planet Earth as speeded up so as to occupy one 365-day year. For the first four months, there is no life. There are no birds or mammals until mid-December. Shortly after ten o'clock on New Year's Eve, there is proto-man. Recorded human history would be lost in the initial whir of the clock striking midnight. Such has been the time for strutting and fretting allotted to the poor player by the master script. Shakespeare, in one of his numerous misanthropic moments, concluded that one lifetime of sound and fury signified nothing. No doubt, had he known biology, he would have said the same for the lifetime of a species.

Sometimes it is useful to think of single organisms as groups of organisms, and vice versa. A human being is made up of millions upon millions of individual cells, so a human being is a community of cells. A termite hill, comprised of thousands of individual termites, can be thought of as one animal with its component cells. Cells are forever dying and being replaced – as are individuals and species. Yet many persons continue to maintain "the immortality of man."

Like a star, or a planet, a species is an individual being in that it is born, it lives, and it disappears. I am not an exact copy of my father, or my grandfather; and my sons and I are not identical. I evolved from other men who looked different from me, and other men evolve from me, but not in my image. Modern man evolved from an earlier kind of man who also looked different; and the lifetime of both, as species, is limited.

My personal extinction comes with my death; and since people make it their business to care about such things, the date and even the hour of my leave-taking may well be known. But the lifetime of a species is so long that its death (as well as its birth) is usually blurred, and it is often very difficult to determine the precise point at which it disappeared. That is because there are two ways of becoming extinct.

We know the date (September 1, 1914) of the death of the last passenger pigeon, for the species was wiped out abruptly and the last individual perished in a cage. We shall never know, however, the date of the passing of the proto-passenger pigeon, for it evolved into something else. In one sense it never became extinct; it lived on in the passenger pigeon which it had become. On the other hand, one could claim that the proto-passenger pigeon became extinct when the new passenger pigeon became an entity. Whether *Homo erectus* still lives, in slightly revised form, we do not yet know; but we are fairly sure that a kind of *Australopithecus* still exists, albeit in substantially altered form.

The same considerations apply to any species, present or extinct. I was there — and so were you — and so were the dinosaur and the sabre-tooth and the Neanderthal, when Jehovah released the first lightning bolt into the primal broth. It has been a long continuum, and it is not yet over. Some of us are still here physically, and all of us always were here genetically. And those of us who are here now — hummingbirds and dandelions and people and squids and damsel flies — have an unbroken genealogy that goes back to the amino acid soup. We were all in it together. We still are.

The fossil record of past ages of life is essentially a history of extinctions, for none of the ancient creatures are alive today — at least, not in the same form. The fossil record is also the history of an uncountable number of experiments. Each new living thing — each individual plant or animal — is an experiment. Each is an "attempt" to probe farther into the possibilities of the biosphere. Like human fingerprints or zebra stripes or giraffe blotches, no two genetic patterns are identical. Each new individual differs from its progenitor.

By mixing the genetic inheritances of two individuals, sexual reproduction creates a unique being. Random genetic fluctuations in the offspring further ensure that change — the one constant in all of life — continues. Garrett Hardin describes heredity as "fate's lottery," and the way in which chance enters into the fate of each new individual as "The Great Fertilization Sweepstakes."

Many persons find the element of chance in life distasteful, even repellent. However, appreciation of the enormous odds involved in the lottery which is human conception should make us even more aware of our individual uniqueness.

Chance is the element which keeps everything else going, whether we like to admit it or not.

Genetic change is in fact change for the sake of change. Animal and plant species must change in order to survive in changing environments. Over the history of the planet there have been constant climatic and geological transformations. Wet hot periods and dry hot periods, frozen periods and temperate periods, have followed upon each other since the world began. There have been times of mountain-building, flooding, and continental drift.

In the course of this never-ending flux, plants and animals of various kinds have encountered a bewildering variety of circumstances. Reptiles can tolerate neither intense heat nor intense cold, because of their inability to regulate their internal temperature, and the peaks and hollows in reptile fortunes have been conspicuous. Mammals and birds, on the other hand, with their homeostatic warm blood control, can venture anywhere in any kind of weather. They can survive from the equator to the pole, so long as there is food and shelter. Over the long haul, birds and mammals have been more "successful" than reptiles. They have penetrated to every corner of the globe.

Evolution involves perpetual exploration of new environmental opportunities. Life as a whole may be described as "opportunistic." As in reproduction, the tendency is to press against the extreme limits of possibility. One might picture a lichen, or perhaps some small vine, as representative of life. With its newest shoots or tendrils it is forever exploring new cracks, new crevices and niches in which it may fasten the tip of a searching growth-bud and there perpetuate itself.

Birds provide ready examples of the opportunistic exploratory process. From the nucleic bird, whatever it may have been like (something had to precede the feathered *Archaeopteryx*), a million different kinds of birds evolved, of which some 8,600 are on Earth at the moment. Each of these species represents the green searching tip of a tendril, always in readiness for new opportunities.

The growth tip of one of the 8,600 bird tendrils is the green heron. There is in South America a very similar and closely related species, the streaked heron, representing the other half of the last fork in that particular tendril. Now, however, the green heron shoot has turned another unexpected corner and encountered a new and empty niche, and it is no longer a green heron. It is sooty black. It has found itself in the Galapagos Islands, and there it has evolved into something new – a lava heron. The green heron on the mainland has remained the same, but it has produced a new species before our very eyes.

The Galapagos, an archipelago in the tropical Pacific, has made an inestimable contribution to our understanding of the processes of evolution. Oceanic islands, removed from the mainstream of life on large continental land masses, have often gone their own peculiar ways. The Hawaiian islands of the Pacific, the Mascarenes, Seychelles and Aldabra of the Indian Ocean, and many others, have all produced unique faunas. But Galapagos is supreme as the classic instance of the appearance of strange endemic wildlife populations. Galapagos iguanas, tortoises, lava lizards, gulls, and finches, are world-famous.

Isolation from continental biological influences allows oceanic islands to give rise to new and unique beings. In the case of Galapagos, which lies on the equator about 600 miles west of the coast of Ecuador, the distance is sufficient to prevent many of the more sedentary forms of life from intermingling with its inhabitants, and this has hastened the development of a new being from something very old.

The beauty of Galapagos is not its age, but its *youth*. Some of the islands are no older than the genus *Homo*, which is very young. By comparison with the South American mainland, Galapagos was born yesterday.

The islands are the individual summits of immense submarine volcanoes which have risen from the floor of the sea some two miles below. Current understanding of geophysical phenomena would indicate that these volcanoes (some of which are still highly active) are manifestations of ocean bed movements associated with the drift of continents.

Once risen from the sea, the islands were nothing but bleak and barren cones of hardened lava. The rocks were bare and lifeless. Eventually, prevailing winds brought the minute spores of pioneering lichens. Due to their fungus component, the lichens were able to take a foothold, and to spread. (The islands today are a lichenologist's paradise. One expert spent a full season on one island, and such was the profusion and variety of his favourite plants that he was able to work his way only a quarter of a mile inland.)

The work of lichens is remarkably slow, but they have ample time and they are the hardiest of plants. They may live for thousands of years. Gradually they break down their rocky bases into minute soil fragments which over the years accumulate in cracks and crevices. Thus the way is prepared for the adventitious arrival on the wind of small-seeded plants such as grasses. Erosion helps the process, and eventually the volcanic rock is broken down into a thin and still tentative covering of soil, in which larger and more elaborate plants can become established. Also, the presence of grasses and other plants may then make it possible for plant-eating animals, such as insects, to survive if they happen to be swept on some weather system to the islands.

At any given moment there must be tremendous numbers of insects and other small invertebrates being carried from place to place on the winds. Most of those which find themselves over the oceans will perish. A few, however, might be deposited on some oceanic island. If they find no suitable food, they too will die.

If the insect happens to arrive after its favourite food plant has become established, then it may well survive. The environment has allowed it to pass through the "selective screen." This, it appears, is the way animals first became established on islands such as Galapagos, and it illustrates the screening or filtering process which Darwin described as "natural selection."

The same mechanism operates on larger and more complicated kinds of animals, whose needs may be just as specialized. There are two species of flycatchers in Galapagos. A wandering, wind-buffeted flycatcher could not have survived upon arrival in the islands if there had not already been an established population of flying insects to sustain it. In the flycatchers' case, the selective screen was represented by the presence or absence of flies. Other animals, such as lizards and snakes, which could not be carried on the winds, no doubt arrived in Galapagos on rafts of flotsam which drifted on prevailing currents from the mainland. In the case of iguanas, the selective screen would be the presence or absence of certain kinds of vegetation, such as cactus. The snake would need a supply of small animals. One way or another, there was an orderly sequence of events which allowed the islands to be invaded and subsequently populated by gliders, fliers, drifters and seafarers from the Americas.

It is probable that only a few individuals of each species arrived – perhaps half a dozen tropical mockingbirds, one pregnant bat, one durable egg-case of an invertebrate without any adult being at all. But the small number of invaders is significant in an evolutionary sense. Those that survived long enough to reproduce passed on their peculiar characteristics. There were so few of them that Galapagos wildlife populations began with a heavily biased sample of the genetic potential of the species concerned. Sheer chance had brought certain individuals, and since each individual is genetically different from all others of a species, the invaders represented peculiarities not common to their species as a whole.

Animals of all kinds have a tendency to throw off "sports" or "mutants" from time to time. In a brood of young animals, there may be one which lacks pigment – a white one, or albino. Usually these do not survive for very long. They are either so conspicuous that they represent an easy target for predators, or they may not be accepted into breeding populations of their species, and be screened out socially. Even if they find mates, their albinism is usually "bred out" in subsequent generations.

In the case of melanism, an animal receives a larger quantity of dark pigment in its inheritance. The black panther is a melanistic leopard; the black squirrel is a melanistic version of the grey squirrel. In many species, these dark "sports" survive to lead normal lives. Their black colour may, in fact, serve as camouflage. It matters little what colour a night-hunting leopard may be. For the little heron in Galapagos, the black lava flows might serve as excellent concealment. The heron a little darker than the others might live longer, breed in more successive seasons, and thus have more survivors. Most important, it would pass on the darkness which allowed it to live longer. Gradually, the number of dark herons would become greater and greater, and the number of more highly coloured individuals would drop proportionately. The natural selectivity of the environment, in this case represented by lava rocks, would have chosen the survivors on the basis of colour. A new kind of heron would emerge.

As the Galapagos Islands age, as more and more of the rocks are reduced to soil, and as the soil gradually permits the development of more green plants on the barren areas, it may no longer be advantageous for a heron to be black. The environment may then be selecting not dark birds but green ones — "throw-backs" to the original form which will return through the generations. More of these will survive than are presently permitted, and a majority of the population will again be more colourful. The dark birds will return to the minority, and perhaps for a while disappear altogether.

The significance of the Galapagos archipelago does not end with the fact that its animals have become different from their opposite numbers (and antecedents) on the mainland. Things have gone a great deal further than that. A new mutation such as colour, which may be beneficial to an animal on the Galapagos, would probably have been suppressed on the mainland. On some neighbouring island a mutation may have gone in a different direction. Given sufficient time, and the fact that animals such as small lizards are not successful swimmers, the populations on two or more islands may begin to differ from each other, as well as from their mainland ancestors. They become different animals. Eventually, they become so different that even if they were united, they would no longer be capable of interbreeding. That is our concept of a species. The operative phrase is "reproductive isolation."

Species are plastic and consist of many varieties. The song sparrow of North America includes more than thirty different geographic varieties (races, or sub-species) of the bird. Each is potentially a new species. The point strikes home when you arrange on a table a number of skins of song sparrows from a museum collection. You might arrange them by colour — from the palest to the darkest – or by size, or by bill characteristics. But whichever way you arrange them, you are struck by the fact that no two of them are *exactly* the same. If your life depended upon your ability to select one and say, "This is the song sparrow," you could not do it. You could only say, "This is *a* song sparrow."

Regardless of their variation, they all maintain their integrity as song sparrows. Each is a genetic experiment, and if one of them should be isolated long enough it would undoubtedly evolve into a distinctly new bird – no longer a song sparrow, but something else. In the Galapagos there are now four kinds of mockingbirds where once there was one. No doubt one day there will be new kinds of song sparrows. So "species" is not, in fact, a reality. It is merely a convenient description of a point in time. The true reality is in the individual animal, with his genetic inheritance, claims the famous zoologist Ernst Mayr; and no individual can be identical to any other. It is the latest in the unending series of chance genetic experiments.

There are in Canada two kinds of flickers – rather large, brown woodpeckers which do a good deal of foraging on the ground. The eastern form has yellow feathers in its wings and tail; it is called the yellow-shafted flicker. The western bird is red-shafted. The eastern form lives from the Maritimes to the Yukon, but not in British Columbia. There it is replaced by the red-shafted bird.

Most authorities recognize these two as distinct and different species. Yet they commonly "hybridize" where their ranges overlap in Alberta. There, resulting offspring may be either yellow-shafted or red-shafted, or a mixture of the two. It would seem that the two flickers have not yet diverged sufficiently; and were they to unite, their "specific" distinctions might quickly disappear.

There are in the world today about 8,600 distinct species of birds. Of mammals there are about 5,000, of reptiles about 6,000, of amphibians some 1,5000, and of fishes something in excess of 20,000. Thus the existing number of vertebrate species is something over 40,000. Of which man is one. Forty thousand green searching tendril-tips in the vertebrates alone. Numbers of species of invertebrates are astronomical; there may be more than a million different kinds of insects, to say nothing of their innumerable relatives. We humans represent only one experiment among millions.

No species is "more advanced" than any of its predecessors. We live in an arbitrary moment on a long continuum, and species existing today are those which have in their genetic make-up the attributes for survival in the various world environments as they chance to exist at this moment. For an animal to have been extant at any given instant over the ages means that it had in its genetic make-up at that time the necessary attributes for survival, and its environment allowed it to survive.

"Survival of the fittest" is the survival of those animals best equipped — best adapted — in the context of their environment at a given time and in a given place. But since no environment is static, individuals, species, groups of species and whole faunas have found themselves maladapted to new situations and have given way to other animals which, quite by chance, found themselves the possessors of a genetic inheritance which was suitable to the new conditions which arose.

The new lava-coloured "green" heron of the Galapagos represents a classic type of adaptation. The rocky environment favoured natural selective screening for colour. Adaptation can take many forms; it can involve the physiology and even the behaviour of an animal in addition to its anatomy and morphology. Because we are such close mammalian relatives, the giraffe has the same number of vertebrae in its neck as we have, but it has taken a different form. The African bush environment has allowed long-necked and long-tongued giraffes to make a very successful living from browsing on acacia trees.

One of the penguin's most obvious characteristics is its inability to fly. If circumstances should change, the bird which has "abrogated" the power of flight could be seriously endangered. But in the antarctic, penguins have no land-based predators, and they are fairly safe in the water. A few fall prey to leopard seals, but not enough to make any difference to the population as a whole.

It is not entirely true, however, to say that a penguin cannot fly. It can no longer fly in the *air*, but it does fly in the water. Penguin wings have evolved into stiff, sturdy flippers, and penguins are as graceful underwater as fishes or seals. They are descended from flying birds, but their way of life no longer depends on aerial prowess. They have relinquished air flight for mastery of the water. In the same sense a seal or a sea lion has given up the mobility on land which once was the property of its doglike or bearlike ancestor in exchange for a suppleness and elegance in the water which is unmatched in the animal kingdom, save by the penguins.

These changes were not "purposeful." The seal and the penguin did not "decide" to re-invade the water which gave rise to their earliest ancestors. They clearly liked the water, as the ducks and grebes and otters do today. Chance mutations which made them even more adept at handling themselves in the water were allowed to persist because they increased the animals' efficiency. (A sea otter is better adapted, and thus more committed, to the water than a river otter or a mink.) As they became more and more specialized for swimming, they became less agile on land.

The world-renowned vertebrate paleontologist and teacher, Alfred S. Romer, explains the sequence of events which changed the ancient crossopterygian (lobe-finned) fishes into the world's first land animals, the amphibians. The Devonian period, the Age of Fishes, is thought to have seen marked changes in world climate, and thus in local environments. At some time, widespread drought is thought to have gradually dried up great numbers of ponds, lakes, and marshes. This was hard on the fishes. Some of them, however, could breathe air, and the lobe-fins also had thick, fleshy fins almost like little limbs.

As these strange, stumpy-finned fishes wriggled, struggled and crawled over the mud, they were making their way overland in search of water. Gradually their swollen, sturdy fins became stronger and eventually evolved into legs. However, they had no notion of remaining on the land, much less of becoming amphibians.

Charles F. Hockett and Robert Ascher in their famous article "The Human Revolution" have expressed this conservatism of adaptation — the force to go on leading the old ways of life under changed conditions — as "Romer's Rule": "The initial survival value of a favorable innovation is conservative, in that it renders possible the maintenance of a traditional way of life in the face of changed circumstances." Eventually the innovation, if successful, becomes one of the means of leading a new way of life in a new kind of situation, in the crossopterygians' case, on the land. Evolutionary conservatism also has implications in the story of human evolution.

A species or a group of species frequently becomes so narrowly adapted to a set of conditions or a way of life that it becomes *over*specialized. A koala bear, for example, eats eucalyptus leaves and nothing else. One conspicuous result of this is the absence of koalas from most of the world's zoos, which cannot supply the fresh eucalyptus leaves. Another possible result would be the total disappearance of koalas, should some blight wipe out eucalyptus trees. The ivory-billed woodpecker, largest of its tribe in North America, was a victim of such overspecialization. It depended on insect food in the oldest bald cypresses of southern swamps. When the swamps were drained and the trees cut, there were no more ivory-billed woodpeckers.

One of the mysteries is the disappearance of the dinosaurs, once the dominant creatures of Earth's land masses. What caused these gigantic beings to vanish almost without any survivors save the crocodilians and that lonely New Zealand waif, the tuatara? Being reptiles, dinosaurs could tolerate ambient temperature variation only within narrow limits. One theory is that the heat became too intense for them: the internal testes of the males heated to the point of sterilization. (The mammalian scrotum is a sperm-cooling device.) Or the world got too cold for them. No doubt the last carnivorous dinosaurs disappeared with their vegetarian fellows, which represented their food supply. The vegetarian dinosaurs may not have been able to endure widespread flooding, or drought. The dinosaurs were survived only by some of the lesser reptiles and by the mammals and birds, which had long since become established.

Great size is itself an overspecialization. It demands immense space and food supply. Another conspicuous over-specialization might be the ludicrously long tail of a male peacock, which must make him extremely vulnerable to predators. Or the apparently unnecessarily wide antlers of a moose. Or the astonishingly large brain of a man. Beyond a certain point, moderation is not only a virtue but an evolutionary necessity.

The most common question about the dinosaurs is "Why?" – not only why they became extinct, but why they got to be so large, and why they evolved in the first place. In nature, asking "Why?" is fruitless. What matters is "How." At any given moment in the long geological continuum, the individuals and species and faunas on Earth are those which because of chance events of climate, geography, and genetics, happened to be the most suited to the particular conditions of that moment. In their time, they *were* the fittest. But times change.

The concept of purpose and direction in evolution as expressed in an anthropocentric pyramid or tree of life is untenable. The world of living things is not a pyramid or a tree or a ladder, with a hierarchy from "primitive" to "highest." Like all worlds, the world of life is a sphere. Each point on its surface is equidistant from its centre. Just as Washington and Hanoi are equidistant from the centre of Earth, the bald eagle, the Vietnam pot-bellied pig, and the human species, are all equidistant from the centre of the sphere of life.

We can reduce the sphere to the two dimensions of a flat lichen community. From its nucleus it expands forever outward toward every point of the compass, just as the cosmos expands into infinity. The concept of direction as expressed in a man-dominated pyramid or tree of life has no relation to reality. To regard the present condition – a monoculture of one dominant species of large mammal – as the inevitable result of a "master plan" is to reveal ignorance and insensitivity. Like other hyper-specializations, the arrogant human brain will eventually be dealt with.

The grandeur of life and life's processes lies in the exquisite opportunism of living things and in the intractable and dispassionate influence of changing environments.

It is sufficient "purpose" for a deadly nightshade vine that its violet and yellow blooms please me in the summer and that its ruby fruits feed some wandering pine grosbeak in the winter.

Beyond that, let us not look for purpose in the cosmos or in life on Earth. It is enough to see, to value – and to savour our capacity for wonder.

SIX

MEASURING *the* UPRIGHT MAN

Through the tortuous courses of evolution, the key factor has been adaptability. The living thing that cannot become adapted to changing circumstances in its environment is screened out. The animal that becomes overspecialized in a certain set of conditions drifts into extinction if those conditions change. A species can be wiped off the face of the planet, or it can evolve into something new, something better equipped to survive the slow but inexorable process of environmental change. Of all the diverse assembly of vertebrates, and the orders of mammals that have overcome varying environmental vicissitudes, among the most successful have been the primates.

A primate is more than a bishop. No doubt it was inevitable that we would ascribe the first order of rank to the mammalian order to which we ourselves belong. How could it be otherwise? In addition to bishops and lesser persons, however, the order of living primates includes such intriguing blood-relations as the tree "shrews" of the Far East, the lemurs, woolly lemurs and aye-aye of Madagascar, the slow loris, pottos and galagos of Africa and the East, the tarsiers of the southwest Pacific, the monkeys and marmosets of Central and South America, and the monkeys, gorillas, gibbons, orang-utan and chimpanzees of Africa and Asia.

Primates appear to have evolved shortly after the midway mark of mammalian history, about 70 million years ago. Scientists have placed primates in a position on the phylogenetic scale immediately following the insectivores and bats and preceding the toothless anteaters, sloths and armadillos, and the pangolins. (Their objectivity in placing the primates there rather than at the end, which is usually regarded as the most "highly" or at least most recently evolved position, is noteworthy, and illustrates why family "trees," with the topmost growth bud inevitably representing man, can be gravely misleading.)

The number and variety of our living primate cousins illuminate our own origins. Over 100 years ago, T.H. Huxley noted that the primates as a group appear to represent a series of more and more "advanced" forms, and that we can infer from them a reasonably accurate picture of the dim evolutionary past. Existing primates enable anatomists to draw some inferences from the increasing number of fossil remains. However, none of the living primates is thought to represent a specific stage in the evolutionary development of the group as a whole.

Of all the characteristics which primates have in common, none is more important than the hand, with opposable thumb. The tree "shrew" does not have this facility, and some believe it should therefore be assigned to the insectivores.

A tree "shrew" has the appearance of a slender little squirrel with a long feathery tail, enormous eyes (it is nocturnal), and a thin pointed nose. The fact that experts are undecided whether this tree-dweller of Malaysia is a primate or an insectivore is illuminating, for it indicates the blurred transitions through which all organisms evolve.

Small mammals were in existence almost since the days of the earliest dinosaurs, but it took time for them to emerge from the dark and tangled security of dense tropical forests. Through the ages of geological time, they became remarkably skillful at surviving, and proliferated into many forms. About 70 million years ago, mammals had diversified to the point that there were in evidence the progenitors of the primates. Judging by existing members of the group, these were probably night-going creatures resembling squirrels and tree "shrews," agile, and expert tree-climbers. They must have eaten whatever vegetable and animal food they could find. Without opposable thumbs, they probably manipulated food with their forelimbs, in much the same way as a chipmunk does today.

These creatures must have required exceptional eyesight in order to judge the distance of leaps from branch to branch. Natural selection for depth perception would undoubtedly be rigorous: it would screen out the small furry creature whose jumps were inaccurate, and favour more rapid movement of the eyes away from the sides of the head, as in a squirrel, and toward the front of the head, as in a monkey.

As the arboreal life became more specialized in its adaptations, some of the primates developed an even wider range of binocular vision and depth perception. Not only were they able to jump with accuracy, they were also able to fiddle with things in their "hands" and see what was going on with both eyes at once. No doubt the fingers and thumb must have begun to emerge at this stage. The hands of modern lemurs give us a clue to this development. Also, improved vision led to advancement in parts of the brain and in its size.

The sense of smell does not appear to have been an important factor to the larger primates. Smaller members of the tribe, such as lemurs and galagos, do rely on it in territorial scent-marking; but the olfactory apparatus has become less and less relevant to the life styles of monkeys, apes, and man. We may *think* that we react with selective sensibility to the smells of flowers, or food, or sex, but these are the coarsest of responses. A dog's world of scent is a dimension which we have totally forgotten.

For millions of years, jungles were full of the earliest primates. But it was long before anything resembling a monkey appeared. By that time, Africa and South America had begun their slow drift away from each other. This may have left some of the earliest proto-primates on each of the drifting chunks. The South American monkeys are vastly different from their Old World counterparts. There is no doubt that they had ancestors in common, but those ancestors must have existed as long ago as 150 to 100 million years. So far as we know, the New World produced no ape-like creatures. The American primates must therefore be exonerated from any hint of implication in the evolution of man.

One of the major criteria for the identification of animals included in the group of newer primates – monkeys, apes, and man – is the reduction in the number of teeth to 32. The oldest forms which thus qualify are dated at approximately 30 million years ago. This seems to have been the time when monkeys and apes began to go their separate ways (man is much more closely related to the great apes than to the monkeys). At the same time, even the ape line was beginning to diverge from the mainstream, and the ancestors of the gibbons were beginning to look like real gibbons. Gibbons, the most acrobatically aerial of the apes, have developed extraordinarily long arms and fingers for swinging from branch to branch. They cannot walk very well, but they run well, with their elongated arms either waving in the air for balance or collapsed and folded turban-like over the head.

Twenty million years ago, before there were apes or monkeys in recognizable form, there were some antecedents. Of these the most important was *Proconsul*, now thought to be an ancestor of the modern gorilla and chimpanzees, but not of ours. Most likely to be directly in our own line is *Ramapithecus*, which has been found in both India and Africa. He is thought to have walked upright.

Uprightness is not necessarily the measure of the man. All kinds of animals can stand on their hind legs, and some are even able to move quickly in that stance. Ostriches and penguins are examples, and in the world of mammals, the bears and a number of primates. Apart from the gibbons, the great apes are not very good standers and walkers, although they are able to when the occasion demands, or when they are taught. The most arboreal of the great apes is the Asian orang-utan, which rarely ventures onto the ground. One young orang in a zoo, trained to stand up on its hind legs, moved slowly and awkwardly with a strange stiff gait, its limbs swinging straight from the hips like those of a peg-legged man. It was christened the "Lock-Knees Monster" by my friend Jim Murray. The poor creature was really not comfortable, but in the gentle good nature so characteristic of orangs, it humoured its audience. Chimps and gorillas can walk a little better, but it appears to be tiring for them.

In his famous paper "Food Transport and the Origin of Hominid Bipedalism" Gordon W. Hewes commented that we "should not overlook the possibility that all cultural systems have a certain vested interest in making the distinctions between man and other animals as sharp as possible. Humanity is self-consciously bipedal." Wilfred Neill, the crocodile authority, states: "Perhaps to man's prehuman ancestors, the first advantage of an upright posture was its effect upon fearsome mammalian predators of that day."

As those who have encountered one know, it is possible to repel, at least momentarily, an angry bull sea lion by maintaining an upright position, which makes you considerably taller than he is. In certain circumstances there is among animals an important aspect about relative size as manifest in *verticality*. The bipedal primate was already well on the way to "sensing" a substantial difference between himself and the rest of the animal kingdom.

How did bipedalism arise? Most primates are creatures of the forests, and beyond the necessity of skipping a few paces along a limb on hind feet while looking for a new hand-hold, the need for walking on two feet would rarely arise. There follows the question: "Why come down out of the trees in the first place?"

Perhaps sheer size and bulk drove our ancestors onto the ground, as it has the largest of contemporary male gorillas who, because of their weight, rarely do much climbing any more. But our ancestors were not comparable in size to the gorillas. Later, there were large species of man-apes, but the primates who forsook the trees for the plains of Africa were believed to be much smaller than we are. Weather is considered to have been the determinant. Environment changed, and various kinds of primates became adapted to the new circumstances.

Sweeping climatic change is said to have broken up the vast stands of tropical African jungle into more or less discrete patches of forest, like islands in an expanding sea of grass. Many of the larger primates remained in their patches of forest, and no doubt in due course they gave rise to today's great apes. Others, however, found it necessary to move out of their shrinking groves and to venture onto the open prairie in search of another suitable woodland. (This is comparable to the lobe-finned fish of the Devonian, in his hour of extremity, struggling to make his way to a better pond.) Many of the primates found new forests in which to pursue the old ways of life; others were incapable of reaching the right kind of woodlands, and remained stranded on the ground. They failed to handle the habitat change in the old, conservative way, and they were forced to handle it another way – through adaptation.

We can picture a form emerging, with slightly shorter fore-limbs and longer legs than a modern ape – perhaps something like a baboon. It moved around on the ground on four legs, occasionally standing up to peer over the grasses for enemies. When it recognized a threat, perhaps it scampered away on its hind legs, managing to keep its head above grass level in order to keep track of the danger.

Bipedalism may have arisen in this way, but the feeling today is that it evolved in response to the need to carry food from place to place. Running on the hind legs released the forelimbs for carrying. (There are those who say that this is abundantly true, but what the little monsters were carrying was not food, but weapons.)

Gordon Hewes believes that the only possible reason for a four-legged primate living on the ground to become a habitual two-footed walker would have been "food transport from the places where food was obtained to a home base where it was consumed." Primate hands are better for carrying than primate jaws are. Jaw-carrying is for cats and dogs, and dog-like primates such as baboons. Even baboons will scuttle along on three legs while holding some tid-bit tucked under the arm like a football. Small forest monkeys such as vervets will often run short distances with booty clasped to their chests.

There is no doubt that our ancestor was substantially a meat-eater. The average primate does not have the right kind of jaws and teeth for cracking bones and slicing tendons, and neither do we. But our ancestor might well have stolen a chunk of meat from the kill of some large carnivore and made off with it to a safe place where it could be picked at leisurely. There must have been many such kills. In those days, the plains of Africa supported an even wider variety of large vegetarian animals than they do today; and there were many predators, such as big cats, dogs, and hyenas.

The little primate, becoming more and more upright, could probably frighten away the jackals just by waving at them. Of course it must have been a different case with the hyenas, which then as now were big, dangerous animals. One can imagine the scavenging primate, in a fit of excitement, shaking a bone, a stick, or even rolling a stone or flinging a piece of turf at the intruding hyena. Slowly our forefather was emerging.

Primates which have become increasingly numerous in museum collections are the australopithecines, and these are closely tied to the genealogical history of man.

Australopithecus existed in many sizes and had equally varied life styles, judging from the fossils. The famous *Zinjanthropus* (now called by its discoverers *Australopithecus*) *boisei*, found at Olduvai Gorge in Tanzania by Louis and Mary Leakey, has been dated at 1.75 million years. What is believed to be the same species was unearthed at Lake Rudolph in Kenya by the Leakeys' son, Richard, in 1970. This one has been dated at 2.6 million years. These were large for australopithecines – they may have been five feet tall. They had immense jaws and grinding teeth twice the size of ours, but their brains were not much bigger than those of gorillas. Their teeth would indicate that, unlike some of the other australopithecines, they were strict vegetarians.

F. Clark Howell describes *Australopithecus* as "a slender four-footer, weighing under 100 pounds." He appears to have been nimble and erect, and, like modern man, he probably ate most everything that came his way, both meat and vegetable. And he had tools. They were mostly pebble "choppers" which had been deliberately flaked. These have been found in great numbers at Olduvai Gorge, together with bones that had been broken to expose the marrow. These tools are relatively "recent" (a couple of million years old). But the use of tools, as opposed to their making, goes back much farther.

The most widely publicized non-primate tool-user is a finch of the Galapagos. In the Galapagos Islands, there are no woodpeckers, tree creepers, nuthatches, or other birds adapted to foraging in the bark of trees. But there is plenty of loose bark in the groves of trees which grow on some of the islands. The bark is filled with the larvae and pupae of insects. In the absence of other birds to do the job, a finch has taken over.

No finch has the special beak of a woodpecker, creeper, or nuthatch, nor the long barbed tongue characteristic of most woodpeckers. So the Galapagos bird must do its foraging some other way. The tool-using finch pecks away at the bark with its bill (which is often open, giving the appearance of striking with the lower mandible) until it finds a food item. Then, unable to reach it with bill or tongue, it flies to a nearby cactus plant and breaks off a spine – or perhaps a twig from some other tree – returns to the site of its activity and pokes around with its small "toothpick" until the prey item is dislodged. Then it drops its probe and snaps up the morsel. Most remarkable of all, perhaps, is the fact that the bird will "fashion" a tool to its liking by breaking a longer twig down to more manageable proportions. Recently the crow of New Caledonia has been observed to probe with sticks in similar fashion.

Another tool-using bird came to light through the observations of Hugo and Jane van Lawick-Goodall. They discovered that the small white Egyptian vulture of East Africa will use stones with which to crack the tough shells of ostrich eggs. (Through experimentation, they also found that the vultures will pick up the eggs of lesser fowl and hurl them to the ground to break them.) One might speculate that gulls, which often drop mussels on rocks to crack them open, might one day decide to drop rocks on mussels.

Tool-using of a different kind is practiced by the handsome satin bower bird of eastern Australia, which uses a tool in the form of a "paint dauber" – a wad of leaves with which it paints the interior of its thatched bower. The colour is a blue pigment made of charcoal.

Only one non-primate mammal is known to use tools. The sea otter of the Pacific coast of North America uses small flat stones to crack shellfish which it brings up from the sea floor. Lolling at the surface on its back, the sea otter places a stone on its chest, then knocks the shells against it until they break open.

The role of tools in the early days of proto-man has perplexed anthropologists. Which came first – the tool, the upright stance, or the burgeoning brain? Which gave rise to which? There are conflicting theories, but they probably developed simultaneously. Environments evolve and animals evolve along with them, with different adaptations developing simultaneously. Probably stance, tools and brain all developed at the same time, constantly feeding back to each other as though by systems "loops," allowing the whole complex being which was to become human to develop at the same relative speed as all of its parts.

Tool-using is different from throwing, which many primates seem to do habitually. One can be showered with sticks and other debris, including feces, released by spider monkeys in the tropical American jungle. An angry chimpanzee in captivity will bowl chunks of turf at its tormentor.

A. Kortlandt, the animal psychologist, has noted remarkable skill and accuracy in throwing by zoo chimpanzees, to the extent that they were able to hit a moving target by aiming in front of it, thus contriving to smash car windows. He has also commented on a young zoo chimpanzee which "attained a hitting accuracy considerably better than that of average children of comparable physical development." Chimpanzees born and raised in captivity have been known to throw "anything at hand" at an aggressive-looking leopard or tiger. It would seem that throwing is deeply ingrained in us primates, if not for purposes of obtaining prey, at least for driving off enemies. A recent film of wild chimpanzees showed the apes taking large sticks with which to attack a dummy leopard suddenly revealed to them.

For maximum throwing skill, however, it helps to be strongly bipedal. Standing on two legs makes it much easier to throw accurately and with force. Any quarterback knows how difficult it is to throw from a sitting position. Only the human primate has produced such throwers as Joe Namath and Dizzy Dean. There is a degree of coordination and balance involved here which is quite impossible for a non-human primate. By comparison with our relatives, some of our physical skills have been vastly underestimated.

The view of Hewes was that true bipedalism arose in this primate as the result of the practice of scavenging meat from the kills of large carnivores, and carrying these bits and pieces away to some home base for subsequent consumption. Tools came later. The opposing school, exemplified in the work of Sherwood L. Washburn and others, holds to the opinion that it was the use of tools which led to the two-legged stance, not the other way round. This is based, quite logically, on the fact that the apes are known to use tools, and that apes are by no means habitually bipedal.

Apes certainly do throw sticks and stones, however, and as Howell puts it, chance success that they may have enjoyed by doing this "may have led to a dawning realization that rocks and clubs were useful as weapons, even for bringing down small game." From there, the theory goes, it was merely a matter of development toward greater efficiency with sticks and stones. This would include carrying the missiles with one, and beginning to run on two legs with them as we know apes and monkeys are capable of doing. The small terrestrial primate was better at it than the monkeys and apes were, and gradually the habitually erect stance evolved – due to the carrying of crude weapons.

Another issue for the paleoanthropologist is whether the stick or rock was in fact a weapon or merely an "implement." Ashley Montagu, the noted anthropologist, maintains that all weapons are tools, but not all tools are weapons. No tool used for killing prey, he feels, is a weapon; the only weapon is the instrument that is used for attacking *one's own species*. A spear, Montagu says, is an implement when it is used to bring down an animal, but it is a weapon when used to strike a man. We must then determine what these early australopithecines were doing. Were they killing other animals, or were they killing each other?

Here the spectre of cannibalism looms – the devouring of one's own species. Cannibalism is *not* the killing and eating of another species, even a closely related one. This is where the "man and nature" syndrome manifests itself: man in one box and "nature" – which means all nonhuman species – in another box. The gentle soul who sees a hawk eating a fowl calls the predator a "cannibal" because it is a bird eating another bird, neglecting to remember that the hawk and the fowl are as different as the gentle soul and the lamb that provided last night's dinner. Cannibalism is rare among birds and mammals, though it is relatively common in cold-blooded animals. It does occur in hawks and owls, where it is in fact fratricide. It occurs occasionally among hyenas and bears, but it is everywhere the exception.

There is little evidence to show that the australopithecines were cannibalistic; indications of it may or may not emerge at some time. But having in mind the general rule, one may be confident that it was rare. The point is, however, that for an *Australopithecus* to make a meal of a *Paranthropus* is no more cannibalistic than it is for a leopard to carry off and eat a Siamese cat, or for a chimp to catch, kill and eat a colobus monkey. They are different species: it is a simple predator-prey relationship. It is probable that there were several australopithecine species, and it is no less probable that from time to time they killed and ate one another. But all the present evidence points to the earlier forms of *Australopithecus* as concentrating on very minor game.

Food gathering and scavenging is a hit-and-miss existence, with the forager settling for what he can find. In total, however, it probably amounts to a well-balanced primate diet. Small lizards, insects and their larvae, birds' eggs and nestlings, the occasional small mammal and whatever chunks of meat could be stolen from carcasses, would nicely supplement the primate mainstay of berries and fruits, nuts, green shoots and tubers. A generalized diet of this kind requires generalized teeth – not the shearing and slashing teeth of dogs and cats, nor the grinding and masticating teeth of ruminants.

Apes (except for gibbons) have notably large, sharp canine teeth, especially in the males. These serve a twofold purpose – for tearing away at the tough covering of much of the vegetable matter on which they depend, and for combat, which includes defence. So far as evolving man is concerned, it is believed that once weapons were developed to a point at which they were more effective than teeth, then natural selection took over and favoured smaller and smaller canines.

There is evidence, however, that the canine began to shrink *before* the australopithecines became habitual tool-users. Even *Ramapithecus*, 14 million years ago, shows signs that the teeth were changing. If, at that stage, teeth were becoming less important, it would indicate that something was replacing the need for them. Perhaps primitive weapons were originating, with cutting implements to reduce meat to bite-size chunks. This seems to have been the case with the australopithecines, and one wonders whether it might have arisen a great deal earlier. If so, the separation of man and nature has far deeper roots than we may have thought.

Both the teeth and the skull of *Australopithecus* had remarkably human characteristics. His upright stance meant that his skull was balanced at the top of his spinal column. He had short, broad hip-bones and substantial buttocks to hold him upright.

The earliest australopithecines very likely ran better than they walked, but their leg bones gradually lengthened, their hips widened, and they became able to slacken the pace somewhat. At no time would they have been able to perform the slow march of the Grenadier Guards without falling over, but they improved, and were able to walk without losing too much of their stability and agility. (It is noteworthy that despite our modern structure, many forest-dwelling peoples today climb trees extraordinarily well.)

It would appear that australopithecines of one kind and another persisted well into the era of men. Where the branching-off took place has yet to be determined, but the earliest of the australopithecines gave rise to both the last of their kind and the first of true men. Whether the earliest *Homo* actually exterminated his blood-relations is not known, but it appears to be highly possible, having in mind that there is no room in nature for the coexistence of competitors. Their ways of life must have been strikingly similar. The two forms, *Homo* and *Australopithecus*, coexisted at least a million years ago in Africa. If one of Richard Leakey's latest finds at Lake Rudolph in Kenya proves to be a prototype of *Homo*, they coexisted over two million years ago.

In 1960, at Olduvai Gorge in Tanzania, the senior Leakeys unearthed a fossil to which they gave the name *Homo habilis*. It was dated at 1.75 million years. Much dispute surrounds this discovery, but certainly it is a far different primate from the australopithecine remains at the same site. Some describe the "handyman" as a transitional stage between the australo-pithecines and the first known man on which there is unanimity, *Homo erectus*. The latter is the same species we used to know variously as *Pithecanthropus erectus* or *Sinanthropus pekinensis*. The "Java man" and the "Peking man" of early texts have turned out to be one and the same, and have been posthumously promoted into the headiest of company, the genus *Homo*. All this underscores the fact that true man has been on the planet much longer than we used to think – in fact, he existed concurrently with the last of the man-apes (or ape-men).

The oldest known example of *Homo erectus* is our ancient acquaintance from Java, who lived more than 700,000 years ago. There is little question that he originated in Africa. His remains have also been discovered from Algeria to South Africa, and from Germany to northern China. He endured for a very long time – so long, that his career overlapped not only that of the later australopithecines but also that of modern man. That he was a transitional form between the two is no longer seriously in doubt; it is now a matter of fitting in the pieces.

Homo erectus was just that — erect man. He stood bolt upright and he was an excellent stroller — so good that even the experts, presented with his leg bones, cannot tell whether they belong to *erectus* or to *sapiens*. The accumulated fossils show that he was about five feet tall, and robustly built. He had a brain about twice the volume of that of the average australopithecine. *Homo erectus'* brain averaged a little less than 1,000 cubic centimetres; but the largest discovered so far reached 1,300 c.c., well into the range occupied by present man.

Erect man used his expanding brain in ways not employed by australopithecines. He explored tool-using further; and before he faded away, his brain had led him to hand-axes, cleavers of fine proportions, and wooden spears. He also recognized the values of fire and advanced that knowledge. He was able to communicate with his fellow man, and had a faint stirring of a cultural tradition.

In view of the growing collections of fossil remains and the apparent overlapping shapes and sizes of skulls, it becomes impossible to assign a precise date to the final disappearance of *Homo erectus*. The last typical specimen is more than 300,000 years old, somewhat coincident with the appearance of men of our own species.

The difference between ourselves and the preceding member of our genus is not in kind; it is in degree. We have smaller features, more bulbous foreheads and less bulging brow-ridges; we are more slender, perhaps somewhat taller. On average, we have larger brains, encased in rounder skulls. We are probably not as strong as he was, but then we do not need to be. All in all, the external difference is not conspicuous.

In the framework of geological or evolutionary time scales, the modern human brain grew to its present dimensions and complexity almost overnight. It evolved far more quickly than the development of our physique, which occurred over millions of years. The brain of a physically highly evolved australopithecine one million years ago was little more than one-third the volume of our own. The brain of an average *Homo erectus* only half a million years ago was on average about two-thirds the volume of our own. Put another way, the brain of the australopithecine virtually doubled in size while it was evolving into the brain of *Homo*. The increase from *erectus* to *sapiens* is smaller. Why this incredible, explosive change from the last man-ape to the first man? It must be attributable to social behaviour. The measure of a primate, after all, even a human primate, is not in what he looks like or what he is or what he says he is. It is in what he *does*.

On PREYING TOGETHER

Most primates are social creatures, and there is little doubt that our pre-human ancestors lived and travelled in bands, as monkeys and apes do today. There is wide variation in the social systems of different kinds of primates, and this seems to be determined by the environmental circumstances within which the animals find themselves rather than by the genealogical relationships of the various species.

One of the more conspicuous behavioral traits of our cousins the gibbons is their noisiness. They whoop up a resounding storm in their dense habitat of tropical Asian jungle. Also noted for their vocal strength are the howler monkeys; big males sound, at any distance, almost like lions or jaguars. But the two are only very distant relations: the gibbon is an Old World ape and the howler is a New World monkey. They are both primates, but here their relationship ends. Both must depend upon their vocal power because they live in thick equatorial forests where it is almost as difficult for sound to penetrate as it is to travel through the vegetation. Keeping in touch with one's immediate group or with other groups is much easier to do vocally than any other way, given the constraints of the environment.

When we were still making a life in the trees, sound communication may not have been as important as it is to the howlers and gibbons. We probably lived at the edge of the woods, in an environment similar to the one baboons inhabit today. But when we became terrestrial, chattering may have helped keep our groups together. (Small birds such as chickadees, warblers, and kinglets, which generally travel in flocks, keep up a perpetual chipping and peeping as though to ensure that everyone knows where everyone else is at a given time, and individuals do not become separated from the main group.)

It would appear that primates group together primarily to protect the younger members of the band. It matters little whether this is conscious or not, for that is the effect of it. One would think that the most important role of sound communication in the earliest days was as warning. A loud alarm note would alert everyone within earshot of danger. Baboons are generally silent, but when one of the group detects a leopard or other interloper in the vicinity, the ensuing "bark" has an uncannily human quality. Such a device must have been employed for millions of years in view of the weakness of primate scent and the sharpness of primate eyesight.

The little primates, proto-men, probably nested in trees at night, as chimpanzees, gorillas, and orang-utans still do. Orangs very rarely come down to the ground, because they are awkward there and the forest floor offers them little inducement for exploration. But chimpanzees spend a great deal of their time there, and the gorillas almost all of it. Yet at sunset they return to the trees, where they build individual sleeping nests wherever they find themselves.

Night-nesting is clearly a response to the ubiquitous presence of a variety of ground-based predators which, with the exception of the leopard, are not good climbers. Even the leopard, though he is reasonably agile in the trees, is not likely to go hunting there. It follows that the heat is off the large apes for a substantial part of every day – half of it, in fact. (Large male gorillas, we are told, do not bother to ascend into the trees. Or rather, they have grown too big to navigate them. They are fairly secure in the nests which they build on the ground. It is hard to imagine a predator in the African jungle with the temerity to take on a male gorilla.)

Observers such as George Schaller have noted that the great apes (in this case, gorillas) climb into their sleeping nests well before darkness falls, and do not leave them until after daybreak. They live in equatorial regions, where day and night are roughly the same length. They sleep for a full twelve hours every day – which would seem to be more rest than they actually need considering that a gorilla's normal daytime routine is anything but active.

One wonders at what point our forebears reduced the twelve hours' sleep to one-third of the 24-hour cycle. Perhaps the development of a high degree of confidence in our survival potential brought about this departure from tradition. Four hours awake in the dark would be unthinkable to a chimpanzee or a gorilla. Perhaps it came with the advent of controllable fire. However it arose, that extra period of daily activity must have had incalculable social consequences. Four extra hours for eating, for tool-making and maintenance, for sex, or for communication. Four extra hours for the conglomeration and cementing of tradition – for learning, for language, and, at some point, for ideas.

It is difficult to conceive of proto-man being sufficiently daring to witness nightfall in the absence of fire. Yet other forms of defence – such as effective weapons or well-developed systems of group action – may have allayed the age-old fears. Whatever the mechanism, the abandonment of the twelve-hour sleeping cycle meant at least some feeling of immunity to natural enemies.

Of all the natural enemies of the primates, the cats are at the head of the list. Lesser cats take their toll of lesser primates, but among the baboons and apes there is no threat like that of the leopard. Tigers occasionally take langurs (long-tailed forest monkeys), but lions rarely if ever stoop to such paltry fare. In addition to the cats, there are the giant snakes, the constricting pythons and others, which take a certain number of primates. Also, there are birds, such as the great monkey-eating eagle of the Philippines. And finally, the primates themselves.

On the basis of present-day primates, fear of darkness seems to be well founded. With a few exceptions – such as the South American night monkey – apes and monkeys do not function after dark. Night is a bad time for a primate, especially a baby one, and it seems obvious why the little primate of our own species demands a night-light in the bedroom, and fears the terrible unknowns of the darkness. One's helplessness is never so conspicuous as it is in darkness.

A substantial case could be made for the existence of a deeply ingrained primate tradition about darkness – and about cats, and snakes, and other threats one cannot see. Our noses long since ceased to be of any defensive value. At night we are left with only our hearing (and our quickening imagination). It is easy to picture, millions of years ago, moonless nights of terror. There have been African nights when I have remembered them.

No ancient fear is more deeply etched in our primate memory than the fear of snakes. A snake does not make himself visible until that last split-second when we are about to walk on him, and then we congeal with terror. It is a long, long memory which goes back to the time we were small animals and snakes were more common than they are today. Usually, by the time you see a snake, the initiative is his. We dislike having our initiative removed, and we especially resent and fear things hidden from our view. The claptrap which attempts to relate women's fear of snakes to their (highly questionable) fear of the male penis is utterly unfounded. Women, who are ground-dwelling primates, have perfectly good reason to be afraid of snakes. So do men.

One would imagine that it was a long time before the proto-man-apes foraged at night. The new primate was a diurnal animal, operating in broad daylight, chiefly by means of his excellent eyesight. From his tree-dwelling ancestors he had inherited the intestinal flora to handle not only vegetable material but also animal protein in various forms. In the early stages, this animal material was probably limited to birds' eggs and nestlings, insect larvae and other minor fare. Later, primates such as the early australopithecines apparently took to scrounging, scavenging and stealing morsels from the kills of large carnivores such as lions and hyenas, and perhaps wild dogs. Lions or their equivalents of the time probably contributed more to the scavengers than did hyenas or dogs; the latter usually bolt their meals so quickly and completely that there is little left for anyone else.

However, a defenceless four-foot primate would be unlikely to walk up to a pride of dozing lions and attempt to make off with a choice cut of their wildebeest. Hyenas, much faster and stronger than the small primate, frequently die in just this way. On the other hand, bands of hyenas are known to be able to drive a lion from its meal. It seems safe to picture a gang of bipedal (perhaps rock-throwing) primates creating sufficient disturbance to distract a lion long enough for their purposes. Young Masai boys today, with a herd of cattle, are equipped to deter lions with only a spear and a long stick.

Eventually, our ancestors learned to kill their own game. No doubt, for ages they had been able to catch certain birds and reptiles, and the slowest of mammals, but this must have been on a very unpredictable basis. It was probably easier to catch middling to large animals than smaller ones. Alone in the woods or on the prairie, it would be virtually impossible to snare a rabbit, or even a mouse, unless one had a well-developed technique. It would be much easier to stalk a herd of antelopes and sneak up on a newly born calf. One may speculate that this was how the killing of larger animals began – by culling out the helpless young and aged. But as soon as australopithecines began to go hunting in parties, the situation changed drastically.

Much attention has been given recently to the social behaviour – including the hunting techniques – of carnivores which hunt in parties, such as lions, wolves, hyenas, and Cape hunting dogs. In their provocative paper, "The Relevance of Carnivore Behavior to the Study of Early Hominids," George B. Schaller and Gordon R. Lowther, while emphasizing that they are not attempting to reconstruct the social life of early man, suggest that "the study of the social carnivores offers numerous possibilities towards the elucidation of the origins and form of social organization in man."

Early man, unlike all other primates, was an organized hunter. It behooves us, therefore, to look to those animals which are social hunters. Thus the *ecology* of early man becomes more important than his primate genealogy. His (genetic) primateness gave man his general physical attributes, but his way of life made him totally different from any of his primate relatives.

There are dangers in big-game hunting, even though the target animal may not seem particularly fearsome. Many large African antelopes have extremely efficient horns and hooves, and even the biggest and strongest carnivores sometimes have difficulties with them. Some of the smaller prey species, too, can be hazardous to tackle. The truculent wart-hog is one of these. The film "Born Free" portrays an inexperienced lioness chasing and bowling over a medium-sized wart-hog which, far from giving up the contest, runs straight at the big cat and strikes her in the ribs.

I once followed a pack of Cape hunting dogs on a late afternoon run over the Serengeti Plain. At considerable risk to Land Rover, photographic equipment, and personnel, we managed to keep up with the smoothly moving skein of dogs until they split up into two groups. I kept up with one bunch, who soon took off at a fast run after a large male wart-hog. After a considerable chase, the pig decided he had had enough running, and turned to face his four or five pursuers. The dogs, apparently perplexed, ringed the boar at a respectful distance and each gave the impression of waiting for another to be the first to risk the razor-honed tushes. A couple of the dogs sat down; one lay down. Finally, what I can only describe in an anthropomorphic way as a fatalistic "shrug" seemed to pass between the dogs, who turned and trotted away. These stand-offs may not be rare. The dogs usually kill by disemboweling their victim, and frequently most of the initial damage is done while the animal is still in full flight, with the dogs nipping and slicing at it from the side and rear. An animal which stands its ground may very often live to run another day. The same holds true with wolves; David Mech says that a bull moose can stand against a pack of fifteen wolves, but if he turns and runs, he will surely be attacked.

The cheetah, which usually hunts alone or in a very small group consisting of a female and her cubs, depends on sheer speed over a limited course. The blinding yellow blur of a cheetah in full run is one of the most stirring and heart-stopping sights a naturalist can experience. The animal begins by stalking a herd of gazelles, periodically "pointing" with a natural elegance no in-bred, pampered hunting dog can remotely approach, until the distance is a quarter mile or less. Then, selecting one individual animal as the target, the cheetah takes off — at first with long smooth bounds which make it very conspicuous (stealth is no longer necessary). Then, gradually quickening the pace, it appears to "shift gears." For the last hundred yards or so, the cheetah moves with amazing speed — at least sixty miles per hour. In the last bound of the chase the cheetah may use its foreleg to trip its quarry up or bowl it over. I have looked at film over and over again in an attempt to see the "moment of truth," but without success. After the cloud of dust settles, the cheetah will be lying beside its prey or over it, apparently giving the *coup de grace* by strangulation.

It is a killing pace, and if the cheetah cannot catch its target within a quarter mile or even substantially less, it will usually give up. Even when the chase has been successful, the cat is almost invariably winded. In the course of one kill, two small cubs which had been left far behind during the long sprint had time to walk up to their mother, open the belly of the dead gazelle and begin feeding before the old animal could get her wind back. (In this case, a hyena came along just a few moments later, drove the cheetahs away, and made off with their meal.)

Lions are even more short-winded than cheetahs, and must make their charges of thirty yards, more or less, from ambush. If they miss at that distance, they must hunt again. (It is surprising how relatively inefficient natural predators are.

Misses are more common than successes, no matter how specialized the predator may be.) Except for loners, who must have a certain amount of difficulty sustaining themselves, lions hunt cooperatively. Most of the killing is done by the females, although males are perfectly capable of catching their own food if they are forced into it.

The lionesses often use a group technique which involves one animal frightening the selected antelope, zebra or other victim toward the others, which are fanned out in hiding. Large cubs are frequently allowed (one might say "encouraged") to participate, although they take a long time to develop their skills to the point of being able to take care of themselves. Sometimes a party of grown cubs will swarm over a large animal such as a buffalo, raking it badly but doing little real damage, until their mothers and aunts finish the process. Lions kill with their jaws, by grabbing an animal by nose or throat, and suffocating it. Indian tigresses have been known to repeatedly bowl over large animals such as water buffalo, apparently so their yearling cubs could get the feel of dealing with large prey; they use much the same killing technique as the lion.

At a lion kill, the adult male or males always eat first, regardless of the fact that they did not usually provide the meal themselves. They are followed – only at their lordships' pleasure – by the females. Finally the cubs are permitted to approach and find what scraps they can, but only after the adults have gorged to the point of exhaustion. Filled with wildebeest or zebra meat, the lions pant wheezily, lying wherever and in whatever position they happened to fall, bellies distended, paws waving limply in the air.

In times of food shortage, such as when the large herds of migrating hoofed animals are absent from their hunting area, the feeding hierarchy of the lions can be hard on small kittens and growing cubs. Then, the smallest members of a pride are so gaunt that their ribs are plainly visible. If the adults are not able to make a substantial kill very soon, these youngsters will fall behind and no doubt be picked up by hyenas or even leopards. This is a drastic but undeniably efficient method of adjusting lion populations to the available food supply.

Undernourishment is not as grave a problem for cheetah or leopard cubs because these animals do not consort together in social groups, and there are no additional mouths to feed after mother and young are satisfied. Also, both species tend to hunt in their immediate vicinity, and the cubs are rarely far from the scene of action. Cape hunting dogs, on the other hand, may run for several miles in the course of a chase. This could make matters difficult for those members of the pack which are unable to keep up, such as small young and their nursing mothers. The dogs demonstrate attractive social grace by bringing food back to their dens. They do not carry pieces of meat in their mouths; they swallow them and regurgitate the meal when they arrive home. Wolves behave similarly. However, no cat is known to do this.

Despite stories to the contrary that one often hears from African farmers and hunters, hunting dogs are friendly, sociable, and gentle. White settlers have long told the most horrendous yarns about the animals (precisely the same sort of fiction one hears about wolves in other parts of the world), always with the emphasis on their ruthless viciousness. There appears to be no basis for these stories. In fact, the van Lawick-Goodalls call hunting dogs "Innocent Killers" in their recent book, which sheds an entirely different light on the animals. (The old-time Canadian newspaperman, Jim Curran, used to say, "Any man who says he's been et by a wolf is a liar.") The word "altruism" was coined by August Compte in 1853; if the capacity did not exist in man prior to that time it most certainly did in wolves and other dogs. If there can be kindness and amiability in the nonhuman world, then both are exemplified in these engaging creatures.

We usually think of killing in terms of aggression against one's own species. When we recognize the sort of killing implicit in predator/prey relationships, we view it in terms of lust, rage, and "animal" ferocity. However, at the moment of attack a carnivore is anything but angry, anything but "fierce." If there is a common aspect to these situations, it is the look of intense *alertness* about the animals. Ears are pricked forward, not laid back as in anger; eyes are wide as in anticipation, not narrowed; tails are raised as though in pleasure, not dropped as in stress; jaws are fairly "smiling." You may see a snarl if you approach a cat's meal or its young, but not during the hunt. One need only watch a house cat hunting butterflies or grasshoppers in the garden, or chasing a ball. He is obviously having fun, and his expression is the same as that of his large relatives at the moment of the kill. At that instant the animals are doing what they do best.

Proto-men and early men were not as swift as the carnivores who shared their African hunting grounds, nor were they as strong or well armed. However, they were organized. Social or group techniques of hunting, such as those of lions, make up for the animals' limited attacking range. Sheer numbers in the case of the relatively small hunting dogs, or wolves, compensate for their comparatively minor armament. Similarly, man-apes and early men could offset their physical limitations with both numbers and organization. The "drive" could be undertaken in a dog-like way, and the ambush could be carried out with the techniques of the lion.

Thus it is that lions and dogs – social carnivores – may tell us more about early man than our nearest living primate relatives are able to.

In their earliest hunting encounters with big game, our ancestors would have been mainly on the watch for animals which were at a disadvantage – an unprotected fawn, or an aged, wounded or otherwise crippled adult. They were probably quick to take advantage of animals mired in mud holes. There comes to mind one aged female Cape buffalo, blind in one eye, which had become stuck up to the withers in a mud wallow. There was no possibility of escape for it. By the next morning, hyenas had consumed the defenceless creature.

George Adamson tells of the periodic droughts which occur in the Lorian Swamp of eastern Kenya. One year, at the centre of the swamp, "there was a pool of semi-liquid mud containing a score or more of rotting carcasses and in the midst a seething mass of dead and dying cat-fish. Further on we came on the bodies of two cow elephants and a little calf embedded in the drying mud. Next to them was a camel cemented into dry mud." His description is reminiscent of the tar pits of Rancho la Brea in California, which have provided a bonanza for the vertebrate paleontologists. There, thousands of creatures of many species became mired in the tar. One would imagine that predators and scavengers could themselves be trapped in such situations, but the elderly buffalo in a small mud wallow presented a perfect opportunity without risk.

For a social animal in the process of developing a remarkable brain, it was probably not a giant leap of the intellect to think of driving prey animals into natural traps such as bogs and soft muddy water holes, or into *cul-de-sacs* where they could be dispatched with stones, large rocks, and eventually spears which may have been sturdy green sticks. The simplest method was to drive herds of animals over cliffs – until recently a standard technique for killing bison in North America. The waste must have been colossal, for the amount of meat which even a large party of hunters could consume or carry away was limited.

Social hunting and scavenging involve sharing of the spoils. The behaviour of lions in their rigidly hierarchical table manners – adult males, adult females, young – certainly cannot be considered social. Their hunting technique is cooperative, but when dealing with the booty, the lions exhibit "selfishness." Schaller and Lowther suggest that the evolution of the social system of lions is "incomplete in that it includes cooperation in hunting but not in the sharing of the prey."

By our standards, a lion kill is mild when compared with the almost orgiastic mayhem of wild dogs and hyenas. When the hyenas make a kill – often on a moonlit night in a group which may number two dozen or more – they tear and bolt their food with an urgency that must be seen (and heard) to be appreciated. Yet, despite the astonishing speed and greediness with which they devour their spoils, the hyenas and dogs do not really squabble over it. Every member of the pack seems to get its share, and they transport food back to those unable to take part in the hunt.

Kills of relatively large game inevitably attract other meat-eaters. Lion kills attract vultures, jackals and hyenas; and hyena kills attract lions. Australopithecine kills must have called attention to themselves in the same way, and no doubt the larger carnivores would frequently drive the small primate hunters from their prey. It follows that the latter would make off with whatever they could well before darkness. Like the hunting dogs, they probably carried food back to their females and young.

It is difficult to determine where *Australopithecus* had his home base. We customarily think in terms of caves. Caves were used by men of the genus *Homo* much later on, but it was probably not the best place for a small primate until he had developed exceptionally good weapons. Perhaps fire and associated amenities preceded cave-dwelling. One might guess that ancient hunting bands congregated at night under an overhang on a rock pile, or in a log-jam of treetrunks overturned by elephants, or perhaps in the trees themselves.

Food-sharing must have eventually contributed to the evolution of work-sharing. The mature males were the active hunters; and the females and young – or at least those females presently with small young – did not participate in the hunt. One can assume that the centrally based females gathered vegetative material and perhaps minor animal food, but the care of their infants prevented them from taking part in big-game hunting.

As the australopithecines grew larger, their young became increasingly helpless; and with the process of evolution, they became more infantile at birth, more "premature." As the size of the skull increased, it became more difficult for it to pass through the female pelvis. The solution was to bear the young at an earlier stage of development, which would then allow the head to move through the pelvis. Prematurity at birth meant longer infancy, and a longer period of infancy resulted in more time in which to learn. In a feedback process, the brain grew faster, and young were increasingly helpless at birth. More energy and time had to be expended in their care, which led to new patterns of social behaviour.

For modern man, growing up is a slow process. We are in the uterus for nine months, we have our first permanent teeth at about 6 years, female reproductive period begins at about 15, and we have our last permanent teeth at about 20. A chimpanzee, by contrast, matures at a dazzling speed: gestation is about the same, but first permanent teeth appear at 3, female reproductivity at 8, and last permanent teeth arrive at about 10. No doubt early man fell somewhere in between. Young proto-men were almost as vulnerable as human infants, which means that the social conditions of their lives must have been elaborately developed for the early communities to survive.

The sharing of the work-load, with the adult males hunting and the adult females guarding their young, suggests that there were un-primatelike sexual conventions by this time. Since these were cooperative hunters, it is unlikely that one dominant male copulated with all females. There may have been proper "pair-bonds" such as we see among birds. These bonds would be able to tolerate the temporary absence of the male hunters. The moment a dominant or "Alpha" male baboon's back is turned, one of his females in oestrus (heat) will be approached by one of his adjutants. For a hunting society to hold together, the baboon way of life would not be appropriate.

The emergence of year-round female sexual receptivity enabled evolving man to deal with the problem of periodic male absences. The female would thus be continually attractive to her mate. In all female primates save man, sexual receptivity occurs only in periods of oestrus, the time of which varies from species to species. When she is not in oestrus, the female does not entertain male overtures, nor is she attractive to the males. The moment she does come on heat, she may be the only individual in that condition at the time, and in many primate species she may mate indiscriminately with as many males as there are in the band, or as many as can get away with it under the watchful eye of "Alpha." (Schaller tells us that gorillas show no sexual jealousy whatever.) In early man, with all the females receptive the year round, it was no longer necessary for the males to watch for the currently receptive female. Lasting sexual bonds could be forged between individual pairs, bonds which were strongly reinforced by the constant presence of infantile young. The monogamous pair with young was now the family. (Carl O. Sauer has suggested that the weakening and loss of the oestrus cycle "is probably a feature of domestication, and it may have occurred early in the history of man, eldest of the domesticated creatures.")

No doubt several such families preyed — and stayed — together. The helplessness of newborn young and the relative helplessness of growing young for periods which now extended into several years meant that defensive mechanisms were well developed by this time. Certainly there was communication of some kind, perhaps "primitive" by our standards, but still the most sophisticated that had yet evolved. Communication does not need to be very elaborate for warning and alarm; in those circumstances the simplest symbol can convey the meaning. But since these were highly social, cooperative hunters, the defence of their home sites must have been organized and effective.

By the time these initial communities had developed, there must have begun to emerge a subtle or even a mildly conspicuous change in the attitudes of proto-men toward their surroundings. Confidence must have been burgeoning. The effective hunter and the effective defender of his family had probably begun to feel a change between himself and the rest of nature. Proto-man could probably not articulate this thought, nor perhaps even entertain it consciously, but the *feeling* was there. The primate place in nature was changing, to a degree never before experienced in the history of any form of life. A creature that could do all the things the australopithecines were doing was also a creature which had sufficient intelligence to sense what was happening — that "nature" was a lessening threat and an increasingly dependable source of the family welfare.

We shall never know when *consciousness* of this change arose. We can only speculate. It is unavoidable, however, to conclude that the social hunter felt increasingly confident about his prowess and about the security of his home base. It is too early to call home base a campsite, probably, because it was informal and very likely shifted position more or less regularly, but it *was* a base. At some stage, the hunters' systems of defence became so efficient that newer and even more un-primatelike modes of behaviour emerged.

One of the most notable of these new forms of behaviour was face-to-face (or belly-to-belly) copulation, in the style sometimes known today as the "missionary position." All primates save man habitually copulate in the usual manner of quadrupeds, with the male approaching from the rear. (The female, if she is not receptive, can simply walk out from under the male.) Some fascinating theories have been advanced regarding the contribution the new sex position may have made to social behaviour in general. Certainly it was convenient for talking – or at least for some kind of oral or facial communication – and presumably it made sex a more ritualized and thus perhaps a more important element of social life than it had been in the past. It may have represented the dawn of affection. Some authors suggest that the female, receiving her mate face to face, might have responded emotionally as she does to the infant at her breast. This may have given rise, according to Hockett and Ascher, to "such further consequences as the Oedipus complex."

Ventral copulation was made possible (but not absolutely necessary) by bipedalism. The buttocks having developed to such a great extent as part of the equipment for an erect stance, and the vagina having shifted from a horizontal to a vertical position as the immediate result of standing upright, in addition to any emotional benefits that may have accrued to it, the new position was simpler mechanically. It is also suggested that the euphemism "sleeping with" rather than "copulating with" arose not in prudery but in the fact that the supine female, with her vagina restored to the ancestral horizontal position, has a better chance of retaining semen than it would if she were standing up or running around. Also, the female orgasm and post-orgasmic relaxation and lethargy could induce her to remain supine for an additional length of time.

More important, belly-to-belly, horizontal sex may well reflect evolving attitudes not only in the human social context but also with surrounding nature. It is difficult to imagine a female primate allowing herself to be pinned on her back by a heavy partner when every surrounding tree or rock might conceal a hungry leopard. Animals simply do not like to be tipped on their backs, and they especially resent being held in that position. It makes them far too vulnerable. Before the new position could become a firm tradition (after all, dorsal copulation is still entirely feasible for man, and widely practiced) there had to be a profound change in attitude. There had to be security from potential predators. The female had to feel especially safe, not only for herself but for her young, before she would have countenanced anything as "unnatural" as this.

It meant a fundamental change in awareness; it might even have been a dawning of the "self" for the individual and, more important, for the species. It shows an indifference to the surroundings which is one of kind rather than of degree. The change was basic and radical. From that point onward, man was different from anything else in nature. Perhaps, in more ways than one, sex made us what we are today. When, as proto-men, we began to enjoy sex in the new position, the leopards did not attack us. They had already learned the meaning of fear.

Exogamy – mating outside the family group or the immediate "tribe" – must have arisen relatively early, if not as a conscious social institution at least as a behavioral pattern. We can be certain that the hunting man-apes knew nothing of incest nor of taboos, but as in other social carnivorean societies such as those of lions, the young animals after adolescence tend to scatter away and to establish prides, or families, of their own. The effect of this is to ensure a proper mix of lion genes, and no group becomes overly inbred. Man, the carnivorous primate, went in the lions' direction rather than remaining in the dominance hierarchy tradition of baboon societies. Washburn, the baboon authority, emphasizes that baboons have the opposite tendency. They breed constantly within their original group, with very little outside mixing. Baboons, of course, are not hunters. The way of life of a species is much more important to its behaviour than its phylogenetic relationships.

Some form of "tacit agreement" seems to underlie exogamy. The mature adults of many species apparently do not want young adults nearby any more than the young adults themselves want to remain close. It appears to suit everybody. It might be hard on the pair-bond if there were many distractions too close to home, and so it would appear that unlike that of nonhuman primate societies, the tendency to splinter off and form new groups was a social factor in earliest times. Initially, the reasons were biological, not cultural. Exogamy arose in a new species of primate living a new kind of life in order to maintain cohesiveness in the structure of a carnivorean family.

Chester S. Chard calls our attention to an additional very interesting point made by Bernard Campbell, who suggests that the very rapid evolution of our forefathers "may prove to be correlated with the appearance of exogamy." Campbell's point is that exogamy would result in a substantial increase in genetic variability "in contrast to inbreeding primates with their slow evolutionary change over millions of years." The wider the mix, the greater the chances for evolutionary change, and the faster the process accelerates.

Assuming that young australopithecines of both sexes drifted away from their families to mate with others of their age group, we must now face the question of relationships between these families and other groups. Did they fight among themselves, or did they share their hunting grounds? Was the pre-human hunter territorial?

The territories which animals defend against others of their species tend to space out populations and to parcel out available food resources in an equitable way, so that there is not too great a strain on those resources in any particular place. Animals vary widely in the extent and intensity of their territoriality, from fiercely competitive little fishes and back-yard birds to the more *laissez faire* attitude of the Cape hunting dogs. There are few general rules about territory in the mammals. Some mammals are vigorously territorial, some are not. As with other social mores, it is more a matter of environmental context and feeding habits than of phyloge-netic relationships.

Thus, in seeking evidence of territoriality in the hunting man-apes, we are advised to look not so much at the behav-iour of present-day primates (which exhibit varying degrees of the custom but among which our closest relatives, the great apes, show little) as to consider the territorial habits of social carnivores. Here, there is equal confusion. The Cape hunting dogs, exemplary practitioners of food-sharing, do not appear to be strongly territorial. They range wherever the hunt takes them, except at the pupping season, when they tend to stay reasonably close to the lactating females with small young. The puppies do not take long to grow sufficiently large to run with the pack, and there would seem to be little need for the dogs to maintain or defend hunting grounds.

Hyenas, on the other hand, *are* territorial. Their hunting methods are much like those of the dogs: a pack of animals chases after a victim until it tires and is torn to pieces. Hyenas do not usually run as far as the dogs do, however, and they invariably return to their dens, which are located in traditional territories. Hans Kruuk, who pioneered studies of spotted hyenas and their behaviour, discovered that in Tanzania's 102-square-mile Ngorongoro Crater the available land was divided up into eight (matriarchal) hyena "clans," and that a clan might contain as many as 100 individuals. The territorial boundaries change from time to time, but at any given time the boundaries of the moment are fiercely defended by the resident packs.

Lions fall somewhere in between the hunting dogs and the hyenas. They do identify territories, but on the other hand they are often very lenient about passers-through, and indeed some individuals and small groups seem to be confirmed wanderers with no home base at all. Wolves, according to David Mech, are probably territorial (they have been observed to attack and drive off lone interlopers), but if they exist as such, the territories are so large that there is little chance of two packs running into each other.

Into which category did the ancient hunter fall? Did he stay put? Probably not, at least until his bands had grown to substantial size. He must have moved around a good deal more than the average lion does, but somewhat less than a hunting dog, having in mind his physical limitations. There is presently no evidence of territoriality in early man, but neither is there anything to refute it.

It would seem reasonable to expect that there is the germ of territoriality in all land mammals. Territoriality in man is probably a "pre-adaptation" — a physical or behavioral characteristic which does not reveal itself until the need for it arises. In man, territoriality would only be manifest in situations such as overpopulation in a given area. Perhaps as numbers of the early hunters grew, there developed in some places traditionally defended core areas and hunting areas.

Territoriality is a phenomenon which exists for the purpose of spacing out animals of the same species, and has no relation to other species which may inhabit the same stretch of landscape. Territoriality exists between man and man, hyena and hyena — not between man and hyena.

Man has been repeatedly charged with "innate aggressiveness." This aggressiveness is conspicuous both in our relations with other people and with other species. It raises the question of dominance, whether it be over one's brother or one's non-human cousin. This is the question of the "killer ape."

In the minds of most people the word "killer" conjures up a vision of diabolical fierceness and cruelty. Our ancestor was not an ape that was a wanton killer, but an ape that was a predator — a hunter — no less amiable than any other ape. Animals which are strongly territorial react strongly to others of their species who invade their personal plots. A male bluebird will vigorously chase away another male bluebird from its wooden fence-post. That does not make the bluebird a fiercely aggressive creature.

We confuse aggression with predation. Predation is a method of food gathering. Is there much difference between picking up a bird's egg or a nestling? Or in taking an adult bird? Or a hare? Or a gazelle? These are only matters of degree. Those who call the pre-human hunter the "killer ape" are not only being misleading, they are offering a dangerous simplification.

We must think of the pre-human hunter as a social carnivore. The fact that he was a primate is incidental. The great apes are remarkable for their gentle dispositions. We can quite properly assume that the last of our forebears who left the trees were not much different in the social sense than the apes we know today.

Once they became social carnivores, our ancestors were like no primates that ever lived before. Their way of life was completely new, for primates. Were the new hunting-and-gathering ape-men dangerous to other animals around them? To their food victims, they certainly were. However, when they were not hunting, the australopithecines were probably just as indifferent to their surroundings as a gorged leopard is today. It is difficult to conceive of a horde of "killer apes" launching ferocious attacks on anything that moved. Predators do not behave that way. Mayhem is a waste of energy, and energy is what the sleeping cat and dog are conserving.

There was no "killer ape," but there is a "killer man." Human aggressiveness has been *learned*. It has been transmitted through uncountable human generations. But it did not originate in the trees. It is not "instinct." It arose in the social and cultural context some time between the day when the first ground-dwelling primate picked up a weapon and the day when man became human. Since aggression is not instinctive, but is learned, we may have hope. Cultures and traditions can change, and they can change rapidly. Perhaps a new culture will abandon our destructive traditions, realizing that these traditions are not "animal." Perhaps we will recognize that we are one with the animals upon which we prey.

EIGHT

FIRE, ICE *and* MEGADEATH

The Pleistocene is generally associated with glaciation. But there is another reason for intense scientific interest in this strange period. The latter part of the Pleistocene (up to about 12,000 or so years ago) saw the extinction, *without replacement*, of large numbers of species of large mammals — all over the world. This was especially conspicuous in North America, where a long list of the very largest mammals became extinct, according to the evidence, at about the same time as man was establishing himself in the "New World."

This phenomenon has been described as "Pleistocene overkill" by Paul S. Martin. The extinctions occurred on a world-wide basis, but the fact that they were so pronounced in North America has led to much questioning and controversy. Extinction has always been the natural way. Extinction is the inevitable future of all species, genera, and even faunas. But extinction has always been gradual, involving the replacement of the extinct forms with new ones. However, the large mammals of the Pleistocene that disappeared did so without descendants and without the emergence of new faunas to succeed them.

Immanuel Velikovsky, the controversial author of such books as *Worlds in Collision* and *Earth in Upheaval*, maintains that there have been cataclysmic upheavals going on in the world all along. These persisted even into historic times, long after the departure of the Pleistocene mammals. Sudden, without warning, and gigantic in scope, these changes engulfed whole populations of animals and brought about the extinction of such creatures as woolly mammoths.

It is true that some of these animals must have met with natural misadventure. Their bodies have been found frozen in the ice, complete with greenstuff in their stomachs. When a bank of the Beresovka River in eastern Siberia collapsed and fell away some years ago, it exposed the fresh-frozen remains of a mastodon. Thawed, it provided "fresh" meat for hungry sledge dogs, which seemed none the worse for downing a meal which had been waiting for them for 20,000 years. Mastodon flesh has even been served at scientific banquets (in very small quantities). The consensus is "only fair." However, one does not need to subscribe to Velikovsky's cataclysmic theories to acknowledge that animals are frequently overtaken by such misadventures as blizzards of unusual intensity. This happens from time to time to bison and domestic livestock on the northern Great Plains during especially severe winters. There is no reason why it should not have happened from time to time to the northern elephants, but it is difficult to accept this as the cause of their total and complete extinction, especially over such a short period of time.

Climate must have played a part in the extinctions of the Pleistocene, as it has through all the ages. But it could not have been climate alone. There was an additional factor, and that was man. It is known that man hunted and killed mastodons and mammoths. For years, naturalists have entertained the notion that some of these great elephants of the Pleistocene may still be wandering the more remote Siberian forests. A more credible thought is raised by one scientist who believes that, in view of the fate of African elephants at the hands of ivory poachers, the last Siberian mammoth may have met its end for the needs of the Chinese ivory market.

The magnitude of Pleistocene overkill was appalling. The record shows an extraordinary acceleration in the rate of extinctions without replacement. Accelerations in the rate of extinctions usually involve some kind of *Homo* in the woodpile. It seems to be a sad fact that when animals disappear on the grand scale (as opposed to the gradual changes of the natural evolutionary scale) man is almost always implicated. Now, human impact on other species moves back into prehistory.

In *Wildlife Crisis*, James Fisher summarized some of the recent extinctions on various continents at about the time of the development of more sophisticated human hunting techniques. In Britain, "man has been a major influence on ecological, faunal and floral changes . . . for 30,000 years at least." In the period between 50,000 and 10,000 years ago, Fisher notes the last British representatives of the cave bear, leopard, forest elephant, sabre-tooth, woolly rhino, lion, and woolly elephant. He assigns culpability for "the demise of some, perhaps most, of the larger mammals" to human hunters.

In the Middle East, a famous deposit of human bone fragments and artifacts, and animal remains, has been uncovered in the Central Jordan Valley near the Sea of Galilee. There are the bones of "some thirty-seven species of mammals, at least thirty of which are extinct." With regard to Africa,

Fisher quotes C.D. Darlington: "In Africa 100,000 years ago men . . . exterminated many genera of great mammals which they killed for food." Martin, proponent of the overkill theory, remarks that despite the wide array of large animals in Africa today, they represent only about 70 per cent of the genera which were there in the middle of the Pleistocene. Stone Age man appears to be inextricably implicated. Simplification of natural systems, including simplification in terms of numbers of kinds of animals, does not happen naturally.

Man is thought to have been in Australia for about 30,000 years. Although Australia's indigenous mammals are all marsupials, there were some very large forms extant at the time of the human invasion of the continent. All of the largest of these are now extinct. There are thought to have been 27 different species of large, flightless moas in New Zealand when the first Polynesians arrived there; none remains today. There are similar accounts from other oceanic islands of the world, and the evidence linking human occupancy to the extinctions of so many of the largest creatures appears to be overwhelming.

The most obvious overkill was that which occurred in North America. Man is a rather recent arrival, in terms of natural chronology. One might think that since man originally evolved on the African plains together with the large animals which still live there, the latter may have been in some kind of equilibrium in a natural way. But the large animals of North America had been totally ignorant of man from their very beginnings, and had had no "conditioning" prior to the invasion from Siberia. It is possible that the arrival of a new kind of predator – especially a technological one – was too much for them to handle. Man critically altered the natural equilibrium. In North America close to one hundred mammals and birds have become extinct – without replacement – in the last 32,000 years or so.

At various times during the Pleistocene there were periods when, because of glaciation, much of Earth's waters were taken out of circulation with the result that sea levels fell substantially. During these periods, the Bering land-bridge was exposed, and there was free movement of animals back and forth, in both directions. At various times horses and camels, which evolved in the Americas, moved westward into Eurasia, and the ancestors of moose, caribou, bison and others moved eastward. With the latter came man.

It has been extremely difficult to determine the timing and the progress of the human invasion of North America. Fossil and archaeological discoveries have been rare, or absent altogether. However, on the gradually accumulating evidence of stone tools found in the Yukon and in Alaska, and across the Bering Strait in Siberia, there has been for some time a credible inference that man came across the land-bridge more than 20,000 years ago. It would now appear that it was much earlier than that.

In 1961, A. Mac S. Stalker, of the Geological Survey of Canada, found human bones on the east bank of the Oldman River near Taber, in southern Alberta. They were the badly broken-up skeletal remains of a child less than two years old. It has been impossible to perform radio-carbon dating on these fragments, because they are too small, but Stalker's 1969 estimate of their age, based on the geological evidence at the site where they were found, is upwards of 37,000 years. In his opinion, "they are considerably older than that, and perhaps as old as 60,000 years."

The investigations of most archaeologists and paleontologists has been drawn considerably farther northward. At various intervals in the glaciation, parts of Alaska, the Yukon, the Mackenzie delta and valley, northeastern British Columbia and northern Alberta were free of ice, and a "corridor" existed which allowed penetration to the interior of the continent. One would assume that these ice-free areas (*refugia*) contained plenty of big-game animals for the late Stone Age hunters. The assumption is that the nomads from the west moved across the top of the mainland to the region of the Mackenzie delta, then southward down the ice-free valley toward the interior. Thus it is possible that, far from being deterred by the last maximum glaciation, men moved south to the Great Plains in the very middle of it, due to the physiography of the time. Indeed it is probable that there were repeated invasions from Siberia over many thousands of years.

We must not think of the human invasion of North America as consisting of a host of marching men, sweeping southward like so many Assyrian cohorts. More likely it was gradual, low in volume, but a trickle which never really stopped. No doubt the hunters followed the movements of the large mammals they depended upon, such as caribou, bison, moose and others. They probably kept up with their seasonal supply of food in much the same way as wolves follow caribou today, or hyenas follow wildebeest.

Some of them, however, give the appearance of having moved rather quickly. Human sites at the extreme southern tip of South America (Tierra del Fuego) have been dated at more than 10,000 years ago. That was a trip of some 11,000 miles from the Alaskan coast, if indeed that is how they got there. There are human dates of 14,000 years more or less from Venezuela, and slightly under 10,000 from Brazil. Fisher remarks, "In the cave faunas of Lagoa Santa in Brazil and in several caves in southern South America is evidence of a vast pampas Pleistocene-type assemblage of big game that could have competed as a show of specialized monsters on even terms with any other fauna of the time, including that of Africa." Certainly this assemblage is not there today, and has not been there in historic times.

A great deal of the evidence in the question of Pleistocene overkill has been circumstantial because of the difficulty of finding human weapons in incontrovertible relationship with the remains of now-extinct large forms of wildlife. The evidence is still being accumulated, but some of it is no longer circumstantial. In the Old Crow area of the Yukon Territory, for example, scientists have discovered bones of the extinct mammoth, broken while fresh, together with human tools, in this case a kind of flesh-scraper fashioned from the leg bone of a caribou or reindeer. Other extinct species found in the same area, together with evidence of human occupancy, were horse, camel, mastodon, ancient bison and a kind of lion, among many others. Radio-carbon dates for the mammoth bones and the fleshing tool coincide at between (roughly) 25,000 and 30,000 years ago.

The Yukon discovery fits neatly with the geological estimate of an ice-free corridor in that part of the north between 35,000 and 25,000 years ago. (Recent evidence of human presence in Mexico some 24,000 years ago, however, together with the southern Alberta material, may push the chronology back considerably farther in the near future.) No one suggests that human invasions may not have taken place in an even earlier ice-free period when the Bering bridge was also present, but the evidence for that has not yet been forthcoming.

Proponents of the Pleistocene overkill theory of large mammal extinction show that the zenith of large game exterminations seems to have occurred between 10,000 and 15,000 years ago, with 11,000 years being the critical period (which, coincidentally, was the date of the beginning of the last retreat of the ice). This is also the rough date of the disappearance of the mammoth, although some did persist somewhat longer than that. The last mastodon may have perished two to three thousand years later; Paul Martin is skeptical of dates more recent than that.

The issue of man's role in the extinction of the great Ice Age mammals is still being vigorously debated. However, it would seem that the disappearance of so many of the largest and most specialized animals concurrently with the period of human penetration into almost all parts of the world is significant. The chain of events probably involved massive kills of the more slow-moving and easily trapped herbivorous animals. Their disappearance made it possible for lesser forms to increase in the vacated habitats, but it also drastically cut down on the numbers of carnivores such as sabre-tooths and giant jaguars which depended on the larger species for food.

A crucial consideration in this investigation involves the time-scales concerned. Because we, in our present form (the Stone Age hunters were men in our present form) are such recent arrivals, we have a tendency to ignore the slowness of biological evolution, as compared to cultural evolution. In *Pleistocene Extinctions: The Search for a Cause*, W.E. Edwards makes this point most persuasively. He emphasizes that for hundreds of thousands of years, the evolution of human culture did not have any appreciable effect on the numbers of animals that men hunted. The australopithecines, and later the "dawn men," were gradually becoming more efficient at hunting, and slowly, almost imperceptibly, they took over from the more traditional scavengers and predators such as the big cats, dogs, hyenas. But — and here is the critical point — "like a population growth curve, the curve of cultural evolution rises exponentially." Once it got under way, it is probable that no adaptation in nature ever evolved so quickly as human culture and hunting efficiency. Biological adaptations were unable to occur and become established fast enough to offset or compensate for the cultural trend. In ancient times, things had progressed in a more or less parallel fashion: as the hunter became swifter, so did the hunted, and as the hunter became more heavily armed, so the hunted became stronger. Not so, however, with evolving culture. The prey species could not keep up with it; genetic mutation is too slow.

Edwards observes, "In the New World the effect was virtually instantaneous." The large faunas of the Americas could not become adapted to cope with a predator that arrived with an already well-developed method of hunting. There was no way for biology to keep pace. There the evidence rests. Those large animals that did survive the Pleistocene overkill were those that then (and now) lived on inaccessible mountains, remote arctic islands, and so forth. Some of them, such as caribou, were those who knew man from time immemorial, and invaded North America with him.

There has been much speculation whether early man in North America (and elsewhere during the same period) made use of fire in driving prey animals to bay. There is little evidence of this prior to the time of the Indians. Although men had had fire, and used it, for hundreds of thousands of years before that time, it is not known to what *extent* they used it. Fire was being used by men long before the appearance of *Homo sapiens*. It has long been known that *Homo erectus* had fire, probably for many, many millennia. In fact, it has been seriously proposed that *Australopithecus* knew and made use of fire, but this evidence is under incendiary debate. We may never know when the use of fire became widespread, much less when the transition was made from the *use* of fire to the *making* of fire, which is quite a different thing.

Many animals are clearly fascinated by fire, and by smoke. And – strangely – by ants. The connection between fire and ants is not as tenuous as it may seem. Anyone who has been bitten by fire ants will know that each nip is as though a tiny burning coal had been pressed against one's skin.

Some animals seem to deliberately invite the attentions of fiery ants. A naturalist knows the peculiar habit of birds which is called "anting." A bird – robin, jay, thrasher, blackbird, oriole, among others – will, on encountering an ant hill, seize individual insects in its bill and insert them at various places in its plumage. It rubs the ants under the wings, over the back and tail, and in all the places one would scrub and sponge while taking a shower. Various explanations for the phenomenon have been offered, such as the possibility that the formic acid of ants may have insecticidal properties and thus help to keep body parasites under control. An alternative suggestion might be that all animals like to scratch and be scratched, and that ants may well produce that satisfaction for birds.

"Anting" sometimes becomes "firing." When birds cannot find ants, they will use such things as cigarette butts, burning matches, hot coals, and smoke. It would seem to be risky, but apparently the attraction outweighs any fear the bird may have of smoldering objects. The fascination with smoke and flame does not stop with birds. Much closer to home, there is in our own primate order the small tarsier of the Philippines which has been named *Tarsier carbonarius* for its habit of "picking up hot embers from campfire sites." Kenneth P. Oakley has observed that "this suggests that man's ancestors far below hominid level of evolution may have been attracted to natural fires and toyed with burning matter." He adds, "After all, is not tobacco-smoking – addiction to which apes and men are equally liable – a form of 'anting' behaviour?"

Our ancestors probably had their initial acquaintance with coals, smoke and flames from natural fires. The Rift Valley of

East Africa, where so many early proto-human remains have been found, is still a "hot" area in a volcanic sense, and may well have been more active several millions of years ago. Undoubtedly, volcanic eruptions were observed from time to time. Presumably also, once their initial alarm had abated, the observers would have been curiously attracted to small, smoldering patches of grass and dry shrubs around the edges of lava flows. And like the "anting" bird or the ember-oriented tarsier, the man-apes would not have been able to resist the new curiosity. Volcanic eruptions, however, are not all that frequent, and one finds it difficult to imagine that this could have been the sole origin of the tradition which, above all others, was to help man become the world's first technical animal.

On a world basis, lightning is much more common than volcanic eruption. About ten thousand thunderstorms take place every day, all over the globe, and each of these generates an average of ten lightning strikes. A total of one hundred thousand lightning bolts per day makes the chance of natural fire from this source a great deal more likely than from volcanoes. Of course much depends on where it strikes. A bolt in the drier parts of the boreal and mixed forest in August is much more likely to start a fire than one which strikes the grassland. Grass has always been thought to be extremely difficult to ignite in this way. However, since our antecedents evolved on the savannahs of Africa, it is there that we should look for lightning fires.

Open grassland was the habitat of so many different species of large grazing animals in Africa which evolved under those conditions that the grass must have been there for millions of years. African people know the values of firing the grass, because it produces fresh green shoots for their domestic animals. Before there were African people to ignite the grass, however, there must have been a sufficient number of natural fires to keep the prairie open. Otherwise, by this time they would have been covered in brush and forest and the grazing animals would long since have vanished.

Grass is extraordinarily resistant to fire; shrubs and trees are not. Something kept those savannahs open for millions of years, in order for the grassland-adapted animals to evolve. No doubt in the earliest times it was due to natural non-human agency. Somewhat later, one might guess that early man or even proto-man might have done his share.

Proof of the use of fire prior to the time of *Homo erectus* may never be forthcoming, but it is believed to have been in widespread use (whether purposefully or not) long before that. Cooking may have arisen quite early, as a form of partial pre-digestion of raw meat. Though our primate innards are quite generalized, they lean more toward the digestion of fruit and other vegetable matter, and easily absorbed material such as eggs, than they do to the long tough fibres of meat. One can speculate that before the "invention" or discovery of cooking, much of the meat consumed was probably a good deal "higher" than even the Englishman's pheasant decomposing on its string.

Fire must have contributed immeasurably to the burgeoning self-confidence of our man-ape ancestors and to the evolving sense of separateness from nature. Many animals use fire in one way or another, and many are not especially afraid of it, but man is the only animal who makes fire and controls it. Fire must have been an important morale-builder, although at no time since he became human did man's psyche need much reinforcement with regard to his relationship with the rest of nature.

There appears to be nothing genetic in the tradition of self-confidence which so quickly evolved into arrogance. Like other primates, man is very low on the "instinctual" scale, and has to learn almost everything he does. The cultural transmission from generation to generation of refinements in tool-making techniques, of improved cooperative hunting methods, and of evolving social conventions of all kinds, was the essential underpinning of the conceptual man/nature dichotomy long before there was *Homo sapiens*, and quite probably before there was *Homo*.

By the time modern man penetrated all the continents during the Pleistocene only 20,000 years ago, his "dominion" over nature was well established.

The POWER and the GLORY

In the early days of modern man, before the carnage of the Pleistocene, weapons were simple and awkward to handle. Yet these crude weapons were capable of inflicting severe damage on even the largest game. One can visualize an exhausted, mired or injured horse, camel, or elephant being slowly crushed by rocks; or bludgeoned, often for a period of hours. One can also imagine sharp sticks being jabbed into a helpless beast, until it died from loss of blood. These proto-spears probably worked in much the same way as the giant fangs of the sabre-toothed cat, which could stab and slash through even the thickest skin of the large animals. No doubt the predator would often begin to eat its prey while it was still alive, as contemporary carnivores frequently do.

Even in those early times, as more sophisticated hunting methods evolved, it must have begun to dawn on early man that he, alone among the animals, was the weapon-wielder, the fire-user, the "dominant."

Although we shall never know when the sense of absolute power became an integral part of human assumption, we do have clues of when the need for power was *acted upon*. The important development was the change from sheer physical striving to *conscious* striving for power over animals. This involved insight, and it is difficult to imagine that men who had insight did not also have a concomitant sense of their status in their immediate habitat. They knew they could perform acts which would have predictable results, and they also knew that the prey animals were not privy to that knowledge. They could plan and execute a hunt much more efficiently than any pack of wolves or pride of lions could do, and they knew it.

Though one could not suggest that in proto-human times there was any human conception of "nature" as apart from man, the earliest *Homo sapiens* undoubtedly recognized his burgeoning power over other creatures. Every success contributed to his swelling confidence, and this in turn "fed back" into increasing efficiency.

Neanderthal practiced cannibalism, and occasionally buried his own dead. He even strewed fresh flowers about the deceased. Elephants and chimpanzees have also been known to place branches over bodies of their kind. This would seem to be a demonstration of affection which is not manifest in the lives of most animals. Neanderthal affection for the departed may be reflection on pleasant things past, or it may involve a form of conceptualizing about the future. One hesitates to speculate that this might include an intellectualization of "life after death," but it does indicate that care for other members of the Neanderthal group extended to the time of death, and — even briefly — beyond.

It is easy for us to say that Neanderthal burials were the first manifestations of consciousness of the finite time-span represented by an individual life. One would expect that this knowledge had long been a part of human culture. Men had seen countless lives — both their own and those of prey animals. The transitory nature of one individual career — whether bear, caribou, or man — must have been well known long before that knowledge was translated into ritualistic behaviour. Other animals appreciate death, and other animals mourn. Neanderthal had been mourning for a long time before he performed the ritual burial.

Neanderthal must have "known" that he was not the same as the animals he preyed upon. Otherwise he would not have buried his dead. No other animal does that. Self-consciousness had begun to show. The difference between one's self and one's prey was so developed by this time that Neanderthal was able to conceptually transpose living animals to animal symbols, thus bringing the animals more and more under his control.

The symbolization of animals was the dawn of magic. It began when the Neanderthal hunter noticed a twisted stump or a peculiar rock formation which quite fortuitously resembled one of the animals he hunted. In conditions of bad visibility, he might even have imagined it *was* the living animal, and then realized the illusion. Perhaps in a moment of light-heartedness he threw a stone, or his spear, at the zoomorphic figure. Neither flight nor fight being forthcoming, the figure became a target for play — or for practice. The hunter had absolute power over the unmoving target; he could strike it at will. In symbolizing the prey animal, he reinforced his power and mastery over the living animal he would hunt tomorrow.

In one Neanderthal cave, deep in an Italian mountain almost 500 yards from its opening, there was discovered a "vaguely zoomorphic" stalagmite at which, it is thought, groups of Neanderthals cast pellets of clay in some ritual which is presumed to have been magical. The cave was littered with the bones of cave bears. Perhaps the stalagmite resembled the living bear, which was then "fired at" in a ceremonious way, either in recollection of past successes or in preparation for new confrontations.

This type of magic is relatively simplistic. So far as we can guess, it did not involve the intervention in the man/bear relationship of any power other than man's own physical and mental capabilities.

The power of the supernatural came by the time of Cro-Magnon man. The great cave-artist was probably the first practitioner of true magic — the deliberate manipulating of natural objects toward the goals of human will.

Many of the beautiful cave paintings are located in remote recesses. They must have been a challenge to even the most agile of cave men. It would seem that for some reason the artist felt it necessary to retire to these deepest cracks, crannies and crevices to perform his ritual. He retired far from the light of day, from the sight of the animals he portrayed, and from his uninitiated contemporaries.

Many of the finest murals consist of paintings super-imposed on one another. In some cases there have been as many as four layers. Why obscure one painting by another? It appears that the magic occurred in the conception of the painting, and each new hunt called for a new one. The places in which superimpositions appear were those which made the best magic.

As the spear is the extension of the arm, so the cave drawing is the extension of the spear. Many of the drawings show animals wounded, with spears protruding from their bodies. Others show animals, such as the mammoth, caught in pits or other traps.

Each graphic representation was an attempt to increase the odds in favour of the hunter. Man and other animals had thus entered into a totally new relationship. Man had ceased to be an integrated part of the natural ecosystem. He now had something which was even more important than flints and spears. In its significance, it ranks only with fire. René Dubos has remarked that the new magic was "probably more important for the understanding of man than physiological and biological knowledge of bodily structure – indeed, more important than the development of tools." From here on, it was the modern era of human dominance.

Whether conscious "propitiation" of wild animals was in evidence by this time is difficult to say. It is well known that men of many societies, before taking an animal's life, have sought its forgiveness in an attempt to ensure that the victim would succumb gracefully, would not injure the hunter and, perhaps more important, would not return to haunt him. Such notions would have to be predicated on the concept of "soul."

Cave art did not stop with hunting magic. The cave people knew that the animals they hunted had to conceive and bear young in order to keep the hunting grounds stocked. Emphasis was given to the fertility of game species. Adults and young were frequently shown. Also, there are illustrations of animals with magnified genitalia and udders, and there are those which are said to suggest the copulation of animals. Cro-Magnon seems to have known the connection between the sex act and birth. He wanted game species to continue to bear offspring, so he used magic not only to secure game but also to ensure the fecundity of game – a Stone Age version of the "sustained yield" concept which only returned to Western human thought after many thousands of years – in fact, in the twentieth century after Christ.

Fertility magic as practiced by Cro-Magnon does not appear to have been restricted to game. Some remarkable figurines have been discovered. These are female statuettes with the secondary sex characteristics – breasts, buttocks, belly – swollen and exaggerated. Apparently Cro-Magnon recognized the need for human fertility as he did for that of the prey he subsisted upon.

Strangely, representations of the human figure are almost entirely restricted to these carvings; they rarely appear in cave paintings. When they do turn up on cave walls, they are usually in the form of relief carvings. Howell remarks, "This is natural, for anyone accustomed to practicing sympathetic magic by making pictures would not be likely to run the mortal risk of having it practiced on him by drawing a picture of himself." Perhaps cave walls were bad magic for people in general. Two dimensions were good enough for animals, but people had to be represented in three! There, it is possible, may be that extra dimension separating man from the rest of nature — flat Cro-Magnon paintings of wild animals, with people in the round.

Although literal or representational pictures of human beings are mostly limited to small sculptures and carvings in relief, there were many cave drawings of *partially* human figures. In these, parts of people were conjoined with parts of animals in a way roughly similar to more recent drawings of mermaids and centaurs. Mermaids and centaurs have human heads and torsos, but Cro-Magnon's animal-men often had humanoid bodies with animal heads and other appendages, Minotaur-like. The Minotaur of Crete was the monstrous get of a supernatural (or, at the very least, superhuman) woman by a bull. But there is no basis on which to impute such decadent fancies to the Cro-Magnon people who, far from being jaded, were very likely much too busy subsisting.

Some believe these animal-men to be the first "gods" of record — the first supernatural beings to have taken literal form in the human imagination. Others suggest they are illustrations of medicine-men summoning up good luck for the hunt tomorrow by putting on the skins and antlers of earlier victims. Perhaps they bounded about in imitation of the living animal's movements. This would be a kind of medicine dance.

These drawings of creatures part human, part animal, may also represent our first excursions into anthropomorphism — the attribution of uniquely human physical characteristics, as well as motives, feelings, and responses, to nonhuman creatures. Anthropomorphism is a tradition which has been dear to man from Cro-Magnon to Walt Disney. It is the result of man's unwavering determination to interpret all things through himself. Nothing can have any meaning unless it is relatable to man; nothing can have any value unless it is possessed of human qualities. If this is true, then Cro-Magnon was very much the modern man. It must be said, however, that he clearly had an awe of and respect for living, sensate, nonhuman life which was largely forgotten in ensuing millennia.

Hunting societies, both ancient and contemporary, have in common their intimate and inseparable harmony with the natural world around them. Dependent as they are on the mammals, birds, fishes, and other components of the landscape, the people are very much aware of seasonal fluctuations in the abundance of certain kinds of food, and thus of the regular rhythms of the year. Hunters are the first naturalists and ecologists. Their lives depend on their knowing and anticipating the cycles of the seasons and the corresponding requirements of the animals they themselves depend upon. As Rolf Edberg says, nature determines the rhythm not only of the hunter's day but also of his years, "filling him with impressions, creating the framework of his emotional experiences." It would be difficult to categorize the emotional experiences of early *Homo sapiens*, but we can be certain that they were compounded of varying degrees of fear, frustration, anxiety, lust, gut satisfaction — and hope. René Dubos has called man "the hoping animal"; that is as good a definition as any.

If early man had hope, his hope was modest and it was usually fulfilled. Even in the most ancient times, man had become accustomed to the turn of the seasons and the regular ebb and flow of populations of animals that were important to him. He learned when the fish would run, when the bears would emerge from hibernation, when the berries would ripen, when the caribou or antelope would come – and go. One would suspect that Cro-Magnon had no special need to spend his hopes on events of this kind; they were predictable. More likely, his hopes were centred on the very short-term – on the daily chase. This was where the magic came in. There is every reason to judge that the original magic, low-intensity though it was, worked well enough. There was plenty of game, and our ancestor had become quite an efficient hunter.

But then there would be the unexpected. A member of the band might be killed in the kind of exploding turmoil that must have been mammoth-hunting. Or a child might be carried off by a hyena. No doubt the hunters were accustomed to such misadventures, or at least inured to them. There were other happenings, however, which could not be anticipated – a difficult winter, or a protracted drought. Accustomed as they were to the more or less regular, cyclic progress of nature, men would be baffled, confused, and frightened by a sudden change in the accepted rhythm of things. One might expect also that some of the natural catastrophes, such as floods, avalanches, and the like, would find their way into the spoken history of the tribe, and thus into its culture. No doubt they would be regarded as "unnatural," perhaps even as manifestations of anti-human vengeance or malevolence.

This was the point at which a far more important kind of propitiation became necessary. It was the time when nature came to be regarded as a force which had to be dealt with, for the good of man. In his formidable *The Golden Bough*, J. G. Frazer tells us that "with the growth of his knowledge, man learns to realize more clearly the vastness of nature and his own littleness and feebleness in presence of it." Recognizing the enormity of the forces, Cro-Magnon's art was only a first step in combating nature.

Eventually the cave artist and sculptor evolved into the full-time shaman. In working his spells on cave walls and ceilings he had demonstrated his "power over nature," and that he could not only draw pictures of animals but of animals in human form. His magic had worked well, but now something much more potent had to be devised.

Frazer says, "In time . . . (man) discovers there are things he cannot do, pains which even the most potent magician is impotent to avoid." Magic must evolve into something better, because these impersonal forces conjured up by the shaman "act automatically and obey whoever applies the right formula." Impersonal forces which could not be properly identified, much less controlled at will, were not sufficient to deal with the larger issues of life. This was the point at which magic began to give way to more personal, homocentric forces aided and abetted by supernatural beings. This was the dawn of religion, and of the evolution from sorcerer to priest.

The role of the priest was to influence the often capricious will of these supernatural beings, by various rituals, familiar incantations, sacrifices, and prayers. The painted anthropomorphic figures on Cro-Magnon cave walls indicate the transition of the painter from magician to sorcerer to priest.

With the advent of religion, man was further alienated from nature. The more human these deities became, the wider the gulf between man and nature. By propitiating and

influencing the gods, man was imposing his will upon surrounding nature and, by extension, upon the cosmos.

Man has spawned an infinite number of religions and
extra-human fancies, but in the matter of the contribution of
religion to man's indifference to nature, none is more interesting than animism. It is customary to link animism with
the pejorative "pagan." Pagan animism was all about souls.

There must be countless definitions of "soul," but if we
accept the most fundamental meaning of the word to be the
life spark, then every individual human being, nonhuman
animal being, and plant being, has a soul. That is incontrovertible. Apparently, Neanderthal man believed it. With his
burial ceremonies he recognized the termination of a unique
being, the departure of life, the departure of a "soul." He came
to realize also that his own life spark would one day leave him.
Perhaps even in Neanderthal times man recognized that other
things – caribou, bears, blueberries – had souls too.

One may describe this type of animism, encompassing all
forms of life, as the reverence preached by Schweitzer.
Animism and Schweitzer's reverence for life are one and the
same. Yet the kind doctor, a devout Protestant Christian,
could never have allied himself with pagan notions. Indeed,
according to his religion, no nonhuman creature could possibly have a soul. There have been many ironies.

Frazer emphasizes a significant transition in the evolution
of religious thought. "When a tree comes to be viewed no
longer as the body of the tree-spirit, but simply as its abode
which it can quit at pleasure . . . animism is passing into
polytheism." He explains that instead of seeing the tree as a
sensate, living being, the man now regards it as "a lifeless, inert
mass, tenanted for a longer or shorter time by a supernatural
being who, as he can pass freely from tree to tree, thereby

enjoys a certain right of possession or lordship over the trees, and, ceasing to be a tree-soul, becomes a forest god."

This is significant in attempting to pin-point the "fall of man." The tree or the deer is no longer a free entity like the man. Man and nature have become entirely different things. The next step, the anthropomorphic one, was to vest the forest god with the body of a man. Man's power over nature was given an irreversible rocket thrust. The loss of animism and the substitution of theism was one of the most critical turning points in history.

It is likely that the domestication of animals and plants was a substantial contributing factor to the elaboration of various supernatural rituals. Fertility of game had been important to man the hunter; fertility of domestic flocks and plantations was of equal importance to man the pastoralist and agriculturist. The earliest dates of domestication are vague, but it seems to have occurred just at the end of the last glaciation some 10 or 11 millennia ago – during or immediately after the peak period of Pleistocene overkill.

We have long fancied that the dog must have been the first domesticated animal, and that man's best friend has been around for as long as we have been capable of interspecific friendship. We have made many speculations about the little motherless wolf puppy brought back to the cave, there to be showered with affection, rescued from the world of savagery and reared in the warm benignity and togetherness of man's world. Man the hunter and dog the hunter forged an inseparable association and relationship to their mutual advantage. The pleasant symbiotic bond has persisted to this day.

There is little evidence for this. Contemporary hunting societies do not use dogs for the chase (Eskimos will loosen their sledge dogs at the critical moment, to bring a bear to bay, but they do not *hunt* with dogs), and there is no reason to believe that early hunting societies did so. If anything, man and wolf are competitors. The likelihood is that they seldom even encountered each other. On those occasions when they did meet, chances are that the wolves behaved much as they do today – they vanished in an instant. Wild dogs do not like human company, and there is no reason to assume that they ever did. Dogs are intelligent, and like big cats and other large mammals, they learned long ago to stay out of man's way.

The more reasonable assumption is that the dog was first domesticated as a source of food. Dogs are still eaten in some societies. There is no "instinctive" man/dog bond. One need only notice how dogs are treated in most parts of the world today, both in non-industrialized and in "developed" societies. I have seen injured and helpless dogs teased, abused and tormented by laughing children in an African desert and in the heart of one of South America's greatest cities. One can imagine that dogs and other domestic animals received no better treatment at the hands of our antecedents.

The sheep appears to have been the first domestic animal. There is a record of a recognizably domestic sheep from 11,000 years ago at Zawi Chemi in Iraq, a mountain site in the east drainage of the Tigris River. Where there were sheep, no doubt there were also goats. Swine are also very old in the history of domestication. These are all quite gentle creatures, and logically would have been the best subjects for domestication. One might guess that the husbandry of these animals, and of dogs, led to the knowledge and ability to deal with large, dangerous beasts such as cattle.

It is possible that cattle were domesticated initially not for meat or even for milk, but as symbols of status. Cattle have a long history in this regard, which persists today in such areas as Masailand in East Africa. A large, ponderous and fierce-looking animal such as a wild bull (or even a cow) would be a conspicuous advertisement of the authority, strength, bravery and thus prestige of the man who owned one. There is also a possibility that even in the earliest times cattle may have had a religious rather than a utilitarian or economic significance.

The domestication of animals was another large feather in the growing warbonnet of the human species. Power over nature had long been associated with magic ritual and the killing of wild prey. The new ability to maintain herds of animals under one's immediate control, subject to one's slightest whim, must have been an enormous boost to the collective ego. The rewards of supremacy could never have been more satisfying than when it was learned that certain animals could be taught to obey human commands.

It has long been thought that the domestication of animals and plants, which we now term the agricultural revolution, began in a relatively confined area in the foothills surrounding the great valley of the Euphrates and the Tigris, and that it spread outward from that nucleus. Recent discoveries indicate that domestication probably did not arise in any one place and that it was not accomplished by any single group of people. According to Chester S. Chard, "Probably each crop and animal developed in a different part of the region and then eventually coalesced to form the standard farming complex when enough alteration had taken place genetically so that each crop could spread beyond its original habitat niche." The first plants to be "tamed" could only be cultivated in the area to which they were native, and it must have been a great many years until artificial selection, whether it was deliberately or unconsciously applied, had produced varieties which could be transplanted into new ecological situations.

With the beginning of agriculture, as with the domestication of animals, the relationship between man and nature entered a new phase. The uncertainty of the chase had been replaced by the certainty of reproduction. The laws of nature were being used in the propagation of captive plants and animals, to be sure, but their very use and the confinement of a variety of living things led to a new type of human social behaviour and a new attitude toward nature as a whole.

By comparison with the time when he had been a hunter and gatherer, man was increasingly independent of external forces (although he soon learned about continuing requirements like soil and water). This had its cultural significance. As J.D. Bernal has described it, the "new kind of society was qualitatively different because of the quantitative increase in the number of people the same land could support." He adds that the transition from hunting to agriculture was the transition we now know in legend as the "fall of man"! Man left Paradise or Eden (the happy hunting ground) to take up working for his bread by the sweat of his brow.

Agriculture meant permanent settlements, which in turn led to what has been called the "institutionalization" of religion in both theory and practice. Power over the animals had become power over the landscape; and since one was now dealing with natural forces in a broader and continuing sense, religious rites became increasingly important in the community. Much attention must have been given to fertility – of animals, of plants, and of people. Religious rites were largely devoted to sex and reproduction, and human sex acts were encouragements for the crops to do likewise, and to be bountiful. The concept of overcoming nature by sheer pressure of numbers would seem to have begun very early; it is no wonder that even in the twentieth century it has been difficult to overcome ancient taboos in connection with the limitation of human numbers.

In the first days of agriculture, embryonic communities were dependent on seasonal rains and flooding. This gave a cyclic, predictable nature to their lives, and was a prominent part of the religions of the time. Prayers and other rituals devoted to rain were not different from the earlier forms of magic, which were still widely practiced in spite of the emerging organized religions. But magic was being rapidly downgraded. Frazer says magic "is now regarded as an encroachment, at once vain and impious, on the domain of the gods, and as such encounters the steady opposition of the priests, whose reputation and influence rise or fall, with those of their gods."

It was possible, however, to help the gods. Even then it was known that the gods mostly help those who help themselves. Artificial supplying of water – irrigation – came quite early, and it united numbers of nuclear communities into cooperative ventures. It was a different kind of agriculture which emerged. A number of villages working on one irrigation project were drawn together into a supervillage, and it was a short step from the supervillage to the city-state. More and more people could live on the land, and the greater number of people meant a new kind of social organization.

It seems likely, on the basis of evidence which has been unearthed in even the earliest communities, that "religion functioned as the original political catalyst in the formation of urban societies and that the early states were strongly theocratic in character, with the priesthood in fact forming the ruling class" (Chard). The progress from priest to king to emperor is well known. And power over men was also power over nature.

TEN

The LADDER of PERFECTION

The conceptual separation between man and nature goes back to the development of tools and weapons, fire and social predation long before the appearance of genus *Homo*. It involves early human magic and cave drawings. Man considered himself apart from nature long before he could consciously articulate the idea, much less before it could be codified, dogmatized, and institutionalized.

Some writers have singled out Christianity as the key factor in modern homocentricity and the origin of all that is worst in our attitudes toward our living environment. It is not as cut-and-dried as that. Christianity, and Judaism, and all other religions and schools of philosophy were vectors or channelers of age-old assumptions – vehicles for the cultural transmission of traditional ideas that have rarely been questioned since the origins of human thought. Nothing biological or cultural remains static. All things evolve and are modified as they evolve, and the churches and schools of thought undoubtedly modified our attitudes. But the germs of the evolution of modern homocentric thought long antedate the Torah, Christianity, and the medieval scholars.

For as long as human societies have existed, people in those societies have attempted to identify a relationship between themselves and the observable facts of the cosmos – the cycles of the seasons and the years, the apparent movement of the sun and the stars, and all the other natural phenomena surrounding them. It was vitally important to explain these things. The mere fact of posing the question, of course, sets man apart from that which he is questioning – the nature he is attempting to relate to. In fact, most such inquiries have been made from the original bias that there are three components to the universe – man, nature, and the supernatural. This fundamental block to productive inquiry has plagued us from the beginning.

Some of the more fanciful cosmogonies are quite attractive, and in moments of light-heartedness one can approve of the healthy skepticism of the Flat Earth Society, for example. One rather likes the old Hindu notion of the giant tortoise on which stand elephants which in turn support the earth. Or the Egyptian picture of the universal egg. One of the Phoenician theories involved spontaneous emanations giving rise to organic matter, to which one can only say, "Why not?"

There is a Polynesian myth of a god of the air who hovers over the waters of the world, breathing life into things. From there it is not a great jump to the Zoroastrian view of a personal deity exercising his own free will in the creation of things; and to the ancient Babylonian creation story, which is so strikingly similar to the opening passages of Genesis. Plato himself conceived of a personal creator. It is the evolution of these anthropomorphic forms and anthropocentric notions to which we must address ourselves in order to comprehend the maturation of man's self-appointed lordship over the planet.

Some of the ancients did in fact conceive of human responsibility in environmental questions. As all conservationists know, Plato bewailed the disappearance of the forests from the slopes of Greek mountains, and noticed that as the result the rainwater ran off directly to the sea without making its contribution to the soil and thus to domestic plants and animals. Clarence J. Glacken in *Man's Role in Changing the Face of the Earth* remarks, "Perhaps if this view had found more elaborate expression in the *Laws*, in which Plato discusses the origin and development of society, an awareness of the philosophical implications of man's activities in changing the environment might have at that time entered the mainstream of Western thought."

Unfortunately, it did not. Mostly, perhaps, because the poet Plato was essentially an idealist and an academic rather than an observer of nature, and although he was clearly aware of the effects of deforestation he was much more interested in his philosophic pursuits. Equally or perhaps even more unfortunate was the failure of the great natural scientist Pythagoras to penetrate and influence the trend of human ideas.

Pythagoras, whose "life-style" was the objective investigation of nature (and who, my colleague C.B. Cragg says, "inadvertently invented Western civilization"), knew that in natural rhythms and relationships there are the "harmonies that all men must seek." As it turned out, Platonism was much too potent for the Pythagorean view of things, and, as Bernal says, conformist Platonism "held back knowledge of the real motion of the heavens, and with it any possibility of valid physics, for 2,000 years." And, of course, man was considered the highest product of creation and it was felt that everything had been created expressly for his purposes. These assumptions were not unique to Greek culture; they can be traced back to some of the earliest Egyptian myths of creation.

Man, the ultimate and highest life form, is unequivocally described by Aristotle. Aristotle's model of the world of nature did not involve either creation or evolution, but it consisted of the familiar scale, starting at the bottom with minerals and soils, then vegetation, then more and more "perfect" animals, with man at the summit. Aristotle took it for granted that there can be no purpose for anything in the world except for human deeds and human artifacts. This notion echoes and re-echoes through human thought, even into the twentieth century, reinforced as it has been by successive dogmas through more than two millennia. It must be said, however, that the generations of churchmen who advanced Aristotle's homocentricity managed to overlook one fact: when he announced that there was no purpose for anything save man, he meant it. There was no purpose for God, either.

In his fascinating *Aristotle*, John Herman Randall, Jr. points out, "The one thing the mature Aristotle did not understand, and apparently had no interest in investigating, was religion. This makes the use of his thought by the great medieval traditions as a religious apologetic seem a colossal irony." Aristotle invested no time in wondering how the world was made, how it evolved, or whether it was created. It was perfectly reasonable in its present form. Man was man, all was there for his examination and comprehension, and that was all that mattered. There was no need for it to have been created.

There was another side to Aristotle. That was his curiosity, and his remarkable insight into the nature of things. This side has been largely forgotten. There is ample evidence that in certain contexts he was uncannily close to ideas which in the twentieth century are considered frontiers of human thought.

Aristotle was essentially a man of science. He recorded things he observed, and he was not concerned with how they got to be that way. Since he did not recognize evolutionary processes, he saw species as immutable; species were more or less possessed of "perfection" according to the various strata in which they found themselves. Aristotle was fascinated, however, by the ways in which different kinds of living things seem to be *fitted* to their distinctive life patterns. This was his observation of the adaptations of animals to their environments; but he saw these adaptations as permanent, fixed, and unchanging.

The Aristotelian world was one with no beginning and no end. Seasons come and go, and repeat themselves; animals are born and in turn give birth, and die, in an unceasing cyclic pattern. The same applied to the cosmos, which was "single and eternal, having no beginning and no end of its whole existence, containing and embracing in itself infinite time."

Significant about Aristotle, however, was that his anthropocentricity was greatly tempered by humility. Randall says, "Aristotle's whole life, as the scholars have now reconstructed it for us, was the realization that intellectual imagination, having plenty of brilliant ideas, is not enough. We need sense also, humility in the face of encountered facts." His unshakable belief – that we must observe, and deal with observed facts, with what actually *is* – endeared him to men such as the deeply religious Charles Darwin, who commented, "Linnaeus and Cuvier have been my two gods; but they were mere schoolboys compared to old Aristotle."

Surely it is one of the intellectual tragedies of all time that Aristotle, in his investigations so incredibly close to a man/nature ethic for his time and ours, was corrupted and perverted by ensuing generations of scholars. Careers have been made of disputing Aristotle. Bernal claims that after Bruno and Galileo, the history of science was largely a story of how Aristotle was overthrown in one field after another. He spoke too soon, and the famous sixteenth-century Frenchman Peter Ramus spoke *much* too soon with his sweeping remark that "everything Aristotle taught was false."

Aristotle knew that the study of observed facts was the proper role for mankind, and that it is pointless to search for mystical origins, supernatural beginnings and ends, when the eternal cycles of nature are everywhere to be seen and comprehended. Irving Babbitt observed that if the champions of the modern spirit who rejected Aristotle "had been more modern they might have seen in him rather a chief precursor. They might have learned from him how to have standards and at the same time not to be immured in dogma."

It is sad that the best of Aristotle was lost and the worst perpetuated. But there is a long human history of original ideas being changed in translation, with the orthodoxy of one age becoming the heresy of the next. By the time of Cicero, man was immovably planted at the centre of the universe. "We are the absolute masters of what the earth produces" by virtue of our position at the summit of nature's ascending scale. Cicero was the thoroughgoing technocrat, and he would have felt very much at home today: "We stop, direct, and turn the rivers: in short, by our hands we endeavor, by our various operations in this world, to make, as it were, another Nature."

Aristotelian worship of the human intellect had its extension in the philosophy of the Stoics, and at one period it seems that there might yet have been hope for a rational philosophy of nature, a spiritual man/nature relationship of the kind that Schweitzer much later proposed. Unfortunately, Stoicism managed to emphasize and underscore the gap between man and the rest of nature by its insistence that man alone has a rational, intelligent soul, and that animals act only by instinct. Intellectual rationalization of man's oneness with the cosmos did not materialize.

Had the spirit of free inquiry into natural phenomena advocated and practiced by Pythagoras and Aristotle been permitted to continue, it would have soon been relatively apparent that man was not in fact at the centre and was a functioning and integrated part of the natural system that was beginning to be understood.

Even more important was the cyclic philosophy of ancient times. Had this been allowed to persist, with its deep and abiding faith in the repeating and immutable nature of things, Judaeo-Christian (and Baconian-Cartesian) "progress" might conceivably have been avoided. The beauty of Greek thinking was its recognition of natural perfection. Things *are*, and they are beautiful. The more we look into and understand the natural sciences, the more we may understand about man. To our misfortune, the monotheistic myths of the Torah, borrowed from ancient Sumeria and Babylon, complete with beginnings and ends, overwhelmed the earlier culture. The new "nonrepetitive and linear" concept of time, completely foreign to the Greek and Roman way of thinking, was to inherit the Western world.

The Torah was said to have been handed down to Moses on Mount Sinai as the Law. The earliest date for the Torah is 444 B.C., two generations before the birth of Aristotle; and there had been a long Hebrew tradition prior to that time. The Book of Genesis is said to deal with the period between 4004 and 1635 B.C., and the injunctions contained therein had long been an integral part of Judaean culture: "Be fruitful, and multiply, and replenish the earth, and subdue it: and have dominion over the fish of the sea, and over the fowl of the air, and over every living thing that moveth upon the earth."

Genesis is a reflection of deep-rooted human convictions, and a component of the basic assumptions that were to be developed in both Jewish and Christian traditions, and thus in Western thought. It postulated a personal God, as Plato had done; it placed man on the uppermost rung of the ladder of living things, as Aristotle had done; it gave the final and ultimate authoritative sanction to man in his role as master of the world.

In his famous essay *The Historical Roots of Our Ecologic Crisis*, Lynn White, Jr. writes: "God planned all of this explicitly for man's benefit and rule; no item in the physical creation had any purpose save to serve man's purposes." It was an ideology easy to accept, for it answered questions that had pestered the best human minds since the beginning: "Who am I? Why am I here? What is my place?" Man, in the image of a personal and all-powerful and benevolent deity, was given mastery of all that was nonhuman. Man had an immortal soul. Non-man had no immortal soul. All things were caused and created to benefit the earthly image of an anthropomorphic God. To this was added Christianity.

The first chapter of Genesis is not the whole of it, nor is Genesis the only book in the Bible. As every scholar knows, and as almost every schoolboy knows, there is a potentially infinite number of interpretations which can be made of both Testaments. The "dominion/subjugation" roles of man and nature as spelled out in Genesis are counterbalanced elsewhere by the notion of "stewardship." Many academics and churchmen have called attention to this, and continue to do so. One of the earlier and thus more noteworthy references to this theme is that of Sir Matthew Hale in the seventeenth century who, while acknowledging that man was at the summit of the (Aristotelian) scale, still felt that he had concomitant responsibilities, for man is a caretaker, a "steward of God." It probably need not be added that this was an aberrant posture for his time.

The idea of "stewardship" is this: God has given man the planet Earth and everything on and in it, but God has also given man the responsibility to keep all these things in as good condition as they were when he received them. Man is God's steward. This is undeniably present in the Bible, but it is no whit different in its anthropocentricity than any of the other "God, man, and nature" references and allegories. Also — and this is even more important in practical terms — it has never been *acted upon*.

Throughout Christian history, the style of Western man's dominion over nature has been that of Genesis 1, and the few sensitive and hopeful souls who have properly recognized the relatively minor theme of stewardship have had no impact on public affairs whatever. The world is full of Christian apologists who call attention to the stewardship theme in defence of their persuasion, but their recognition of it has never progressed beyond the academic stage. It has made no impression upon the "real world." What *has* made an impression is the sense of Genesis 9:2 — "and the fear of thee and the dread of thee shall be upon every living thing. . . ."

Orthodox Christianity is a long, long way from the living world which gives us breath and substance. It revolves first about the existence of one God. The fact that man is made in God's image indicates that this god, like so many of his predecessors, is an anthropomorphic God. He makes it that much more difficult for us to remember man's kinship with nature. Man and God are able to have their private discourse, from which nature is excluded.

Secondly, there is the concept of "original sin," and the "fall of man." This is tantamount to the "killer ape" theory. Our origins are rooted in blood crimes of such appalling enormity that none of us is fit for the "service and fellowship" of God; we are subject to God's displeasure now and in infinite eternity. That there is in fact no "good" or "bad" in nature is ignored or suppressed. If you want to keep them in line, you must keep them frightened.

Finally, since man is so inept, and is incapable of extricating himself from his heritage of wrongdoing and suffering, God sent his son "to save sinners, to deliver them from hell, to make them holy, and partakers of the eternal joy and glory of heaven."

The aggregate of Christian thought is even more man-centred than the Hebrew faith which preceded it. There is no room, anywhere, for man, the extraordinary product of primate evolution. There is no room for man, the glorious artifact of living nature. All is suppressed to the categorized picture of the world, with the whole of nature having been created by God separately from man and for man's (and God's) glory. Whether man is regarded as total and absolute master and consumer, or as steward, does not really matter. The dichotomy is clear. The descending order is God-man-nature. There is no conceivable reason for the existence of the blue planet apart from the needs of God and man.

One of the most elaborate and complex rationalizations of Christianity, nature, and man was presented by Pierre Teilhard de Chardin, who died in 1955. A French Jesuit, he was also a geologist and a paleoanthropologist of distinction, having been involved in the early discoveries and interpretations of *Sinanthropus*, now *Homo erectus*. One's impression of Teilhard is of a brilliant man of science and religious faith in equal measure, and thus a man who was intellectually tortured. He was a philosopher who bent every energy of his immense talent to making "sense" of mankind in the light of twentieth-century scientific knowledge, yet in the sense, also, of the dogma to which he was committed. Perhaps one could say he was dedicated to the reconciliation of faith in man and God with faith in nature and demonstrated natural processes. He was also a mystic. The turns and ramifications of his reasoning are not easy to follow, especially since his style is enormously complicated and he found it necessary to coin words to convey his meanings. But he is worth the effort.

Teilhard recognized the fact of evolution. As a Christian, however, he saw man as "the only absolute parameter of evolution." Evolution existed solely to produce mankind. All of Teilhard's immense intellect was directed, it would seem, toward constructing an anthropogenesis, an explanation for the phenomenon of man. (Having in mind his background, we must accept that an explanation for man *had* to be found; the majestic accident of chance evolutionary process was not enough.)

For Teilhard, as for Aristotle, man was "the unique, terminal, inflorescence on the tree of life." Man's evolutionary development was purposeful; it was toward a defined and absolute goal, "polarized towards a determined point." All of evolution, before the appearance of man, was thus determined; the evolution of man, in its ascending progress, raised the human mind "to the position of constituting a specifically new envelope to the earth." This new envelope, superimposed on the biosphere, was the envelope of *knowing* — the *nöosphere*.

In Teilhard's view, a consciousness pervades everything. In the course of evolution, this consciousness has become more and more highly organized as it has moved up the scale of vegetative and then animal complexity. Always, as with Aristotle, it has moved toward greater and greater perfection. In man, we have the "ultimate and supreme product" of the first stage of evolution.

Man, according to Teilhard, is alone among the animal kingdom in his ability to be conscious of himself, or to reflect. This self-consciousness of man represents the envelope of knowing – the nöosphere.

Teilhard sees the next stage of evolution as the achievement of "collective reflection," or universal knowing – a complete and unified consciousness of all mankind. The goal of this determinist evolutionary progress is an "Omega point." The design is God's, and the achievement of universal consciousness is man's function in it. Teilhard expects that the attainment of collective reflection may require "some millions of years." Ideally and theoretically, "mankind will come to an end when, having finally *understood*, it has, in a total and final reflection, reduced in it everything to a common idea and a common passion."

This is a remarkable statement from a man of science in the twentieth century. It is a demonstration of the infinite capacity of a rarely gifted mind to perform intellectual contortions within the confines of traditional and indisputable dogma. But it is impossible to ignore Teilhard's thesis. Every one of us at times raises the questions, but rarely do we attempt to answer them with the persistence and the elegance – much less the faith – of a Teilhard. His message regarding world-wide understanding is abundantly clear.

Yet, one is obliged to recognize the frightful assumptions which underlie Teilhard's assessment of man's role on Earth. All was placed in readiness by divine plan before the evolutionary "progress" was set in motion, and the master design could only be realized in the form and mind of man. Teilhard says, "Man is irreplaceable. Therefore, however improbable it might seem, he must reach the goal, not necessarily, doubtless, but infallibly." Since there is no reason for the existence of anything on Earth save for the human mind, all else is inconsequential.

We have been entertaining such thoughts, and expressing them, as long as we have had thoughts and the power to express them. The man-centred universe is almost as old as man himself. Our traditions have done their work so well that, as Teilhard discovered, we do not even have the vocabulary to articulate new ideas, much less communicate them.

One would wish that our species had been granted somewhat more humility and somewhat less consciousness of self, or "reflection." Or perhaps one might also wish for more open-mindedness and a great deal less of the capacity for self-deception.

ELEVEN

ZEAL *for* SUBDUAL

The first fifteen centuries of Christianity were a dark and empty void in human awareness of nature. Evolution of objective thought and honest reflection were largely choked off by the medieval church.

The literate minority of people was mostly preoccupied with dogmatic interpretations, often erroneous, of Aristotle and other classical scholars. Scientific and even academic excursions into the living world were not set in motion until the latter part of the period, the dawn of the age of exploration and conquest – both, as it turned out, chiefly in the name of the church. In its final decades, however, the sixteenth century became a kind of intellectual crucible out of which poured the devastating thoughts of Copernicus and Bruno, Galileo and Kepler, who, with the unparalleled Shakespeare, led men's minds into the first fresh air of modern thought.

There is an astonishing gap in the development of natural science between the time of Aristotle and that of Descartes and Bacon. No one was able, or permitted, to pick up where the classical scientists and philosophers had left off. Well before the time of the birth of Christ, philosophy had become something quite different from science and politics, and had taken on an almost entirely moral or ethical tone. Where the philosophers of classical times had been fascinated by the world of nature and the intriguing questions in the observable cosmos, the philosophers of the Dark Ages had become priests; and "science," if it was such, represented what Bernal says might be called the religion of the cultivated upper classes.

The stultifying effect of the medieval churchmen on the growth of human knowledge is epitomized in the fate of astronomy, the most ancient of the sciences. Long before the Christian era, astronomy was known to Chinese, Hindu, Chaldean, Egyptian and Greek philosopher/scientists. Its interpretations varied; all except the Greeks seem to have used their observations of the heavens in the pursuit of the pseudo-scientific aims of astrology. The Chinese made political extrapolations from their observations, while the Hindus, Chaldeans and Egyptians found various ways of building astronomy into their religious structures.

The Greeks made astronomy the first of the exact sciences. Six hundred and forty years before the birth of Christ, Thales, building on the work of the Egyptians, was able to predict a total eclipse of the sun, and to call attention to the value of the constellation *Ursa Minor* (the "Lesser Bear") as an aid to navigation. A practitioner of abstract geometry, Thales could calculate the heights of things and the distances of ships at sea. He was also sufficiently free-thinking to believe that water was the element from which all things had evolved. From his time, through five centuries until the days of the great Hipparchus, who died in 120 B.C., the ancients made extraordinary advances in science, which at about that point were abruptly terminated for fifteen hundred years.

Shortly after Thales, Pythagoras had promulgated the

notion that the earth and other planets move around the sun, but he was not to be recognized for many centuries. It has been said that if these ancient scientists – especially Hipparchus – had had access to the telescope and the pendulum, it would have taken no more than fifty years to bring astronomy to the point at which it was at the time of the birth of Isaac Newton in 1642. Apart from Ptolemy in the second century of Christ, there was no astronomer, and no advance in astronomy, until Copernicus, who died in 1543. He had picked up the motion of the planets where Pythagoras had left it two thousand years before.

It is clear that the Dark Ages were a total loss to the development of human comprehension of the world men live in. The churchmen had kept man firmly and immovably locked at the centre of the universe. The homocentric Christian view of the world was so all-pervasive in Europe, and so unquestioned, that even the few men of science who emerged from time to time during the largely barren centuries were themselves unable to conceive of any structure of the universe other than the Aristotelian ladder of lesser-to-greater perfection, with man securely perched on the topmost rung. And no matter where a man found himself in the hierarchy of society, from peasant to priest, he had no doubt whatever of his personal dominion over nonhuman nature.

Medieval thought was dominated in the thirteenth century by St. Thomas Aquinas, the Italian scholastic philosopher. Aquinas, maintaining that theology and science cannot be mutually exclusive or contradict each other simply because truth is indivisible, set about the enormous task of reconciling the Aristotelian heritage with Christian dogma. As we have seen, the purest of Aristotelian thought does not allow for beginnings and ends, much less for personal Creators, still less for religion of any kind. But the brilliant synthesis of

Aquinas, like the rationale of Teilhard, made the most of Aristotle's ladder of ascending rank, to which was added God. God, man, and nature were rigidly compartmentalized.

The essence of Aquinas' teachings appears to be that there are two sources of knowledge – Christian faith with its attendant mysteries, and the truths arrived at by human reason. In effect, since truth is indivisible, this means that the mysteries of Christian dogma are to be accepted. They need not be understood to benefit us.

Thus, Aquinas managed to set up a body of dogma which allowed symbolic mysteries and mystical analogies to be accepted as part of the eternal truth. Julian Huxley has commented that this results in the analogies and symbolisms "being taken for more than they are – for scientific knowledge, or even for an absolute certainty of some still higher order – and conclusions then drawn from it." These conclusions, Huxley observes, follow with "full syllogistic majesty," but since the premises are false, the exercise is dangerous.

In his brilliant *Civilisation*, Kenneth Clark states that Western civilization could be argued to have been the creation of the church. He notes that he is not thinking, "for the moment, of the Church as the repository of Christian truth and spiritual experience: I am thinking of her, as the twelfth century thought of her, as a power – Ecclesia – sitting like an empress." Of this there is no doubt. The power of the church in medieval times is almost incomprehensible. And the church did not allow for any ethic that extended beyond its dogma.

Schweitzer has remarked on the "medieval contempt for the world and for life," an attitude brought about by the harmonizing of primitive Christianity and Greek metaphysics. "It is fundamentally a rejection of the world and of life because the interest of Christianity at this time concentrated upon otherworldly things." The church was all-powerful, and the church was entirely occupied with the mysteries of its dogma. Schweitzer added that this early form of Christianity also imposed upon people an attitude toward life that was in contradiction to their nature. They endured it, however, for an exceptionally long time.

We can only guess at the thoughts that occupied the mind of a medieval peasant, and we can only guess at his nature, and whether it was being contradicted. No doubt the peasant believed; he had no alternative. He could not read his own language, much less Latin, so he accepted what he was told of divine revelation. No doubt he also believed in the intercession of the saints on his behalf – or at least he realized it could do no harm to propitiate them. He must have dwelt in constant fear of reprisals, here and in eternity. That this fear transferred itself from hellfire in the future to the living presence of priests in the here and now seems certain. Trevelyan tells us of the typical English peasant of the time of Chaucer, who could not follow the Latin liturgies yet was able to absorb a great deal from his surroundings. "Fear of hell was a most potent force, pitilessly exploited by all preachers and confessors, both to enrich the Church and to call sinners to repentance."

There is little way of knowing how medieval people felt about nonhuman nature, or whether they felt anything about it at all. Since people were the unique possessors of immortal souls, however, no doubt the nonhuman elements of the landscape were viewed with mild curiosity, if not contempt. Perhaps it was indifference. One does not need to suppose that medieval man was actively suspicious of nature, but he certainly did see it as alien.

It is interesting how medieval art abounds with pictures of birds and other animals, and of flowers and trees. One had always fancied that this was merely a manifestation of the artist's enthusiasm for embellishment; but there is more to it. The philosophers of the Middle Ages called it *natura naturans* — nature "doing its thing" — flowering, burgeoning, producing. Not far, really, from "pagan" expressions of nature in fertility rites. But, as Clark points out, "The odd thing about the medieval response to nature was that it saw all those things in isolation. It quite literally couldn't see the wood for the trees." That probably explains one's reaction to them as mere decoration; they have no perceptible meaning. Clark adds, "As for birds, they were a medieval obsession. . . . I think the reason is that they had become symbols of freedom." We still use the expression "free as a bird" even though we now know that birds are among the more narrowly "programmed" of animals, with rather little freedom of individual choice built into their high-intensity careers.

Birds make us recall St. Francis of Assisi who in the early years of the thirteenth century preached to the birds, and is said to have tamed a marauding wolf at Gubbio through gentleness. The wolf repented of his ways, "died in the odor of sanctity, and was buried in consecrated ground."

In his understanding and acceptance of the oneness of all living things, and in his comprehension of the value of both individual and collective humility under God, Francis was a rare medieval Christian. As Lynn White, Jr., puts it, "Francis tried to depose man from his monarchy over creation and set up a democracy of all God's creatures." In this, he was doing essentially the same thing as the modern environmentalist or naturalist is doing — trying to create a perspective, in nature, for man as a part of nature. He was unsuccessful.

To accept this interpretation of Francis' teaching, one must reject the notion of man's uniqueness in the possession of immortal soul, and extend the concept of soul to all living things. This was impossible within the confines of church dogma, and this portion of St. Francis' tradition has not survived. His vows of poverty, chastity and obedience have of course been perpetuated in the distinguished religious order which bears his name.

Also out of Italy, with its beginnings about a century after Francis of Assisi, came the Renaissance, and the first new stirrings of "humanism," although it reached full flower two hundred years later. The magnificent art and architecture of the Italian Renaissance were unrestrained celebrations of man. Man was equated with beauty, with joy, with dignity — and with power.

Above all else, it was power. The great Venetian architect of the fifteenth century, Leone Battista Alberti, whom Clark has described as the "quintessential early Renaissance man," expressed it clearly. "A man can do all things, if he will." The men who made such utterances must be judged in the context of their time. The Renaissance was not a period for humility. As artists and scholars gradually rediscovered Greece and Rome, and as they became emancipated from the yoke of medieval religious authority, it was natural that they rejoice in their newly discovered intellectual and artistic freedom. Man was liberated, and his mind was free to assimilate whatever touched him, whatever would contribute to the edifice that he was building — the edifice that was to be the monument to man's intellect.

The Renaissance rediscovered classical literature and classical science. Inquiry into the nature of the world, and its processes, could no longer be suppressed. Developments in mathematics and physics had almost instantaneous results in a practical way. No applied science of the time was to enjoy such immediate benefits as navigation.

There was a neo-classical interest in cartography. By the beginning of the sixteenth century, it became possible to work out longitude. Somewhat later, the emergence of such new hardware as the astrolabe, the sextant, and the quadrant allowed accurate measurement of the position of the sun and the stars, and thus a reasonable indication of the position of oneself. By about 1600 came arithmetic, trigonometry, and the use of logarithms. Thus, to the creative side of the Renaissance was added the extra dimension of sea-faring and exploration. The world was conquerable. No element of the rebirth of knowledge and the resurgence of man's capabilities was to be so important as the susceptibility of Earth to conquest.

In the last decade of the fifteenth century, Bartholomew Diaz had rounded the Cape of Good Hope, Columbus had reached the Bahamas and parts of the West Indies, and Vasco da Gama had made it to India. Vincente Pinzon had discovered the prodigious mouth of the Amazon, Amerigo Vespucci (for whom the new hemisphere would be named) had encountered the Guianas, and John Cabot had reviewed the coastline from Greenland to Virginia. Within twenty years, Magellan's party would circumnavigate the globe, Balboa (not stout Cortes) would cross the isthmus of Panama to the Pacific, and the implacable *conquistadores* of Hernan Cortes would be savagely destroying the ancient and brilliant civilization of the Aztecs in the name of church and king. European man had the entire world in his grasp.

During the great age of exploration, no parts of the world were more vulnerable to conquest than oceanic islands. The physical and biological isolation of such islands and the resultant genetic "wanderings" of their living inhabitants had resulted in some singular evolutionary products. Most important in this regard is the fact that many animals of such islands have let down their defences, as it were. They have become adjusted to a very local and specific set of conditions. They have become completely dependent on a unique way of life.

Many birds of remote oceanic islands have lost the power of flight. It is not necessary to be able to fly if a bird has no land-based enemies. Thus, where there are no land predators, there are the penguins of the Antarctic, the flightless cormorant of the Galapagos, the various flightless rails of New Zealand, Laysan, Tristan da Cunha, and elsewhere. The North Atlantic great auk, now extinct, was also flightless.

All of these are descended from birds which once could fly. But when the need for the power of flight was eliminated — in the form of the absence of natural enemies — natural selection disposed of flight. It is said, in fact, that flight may even be disadvantageous to an island-dwelling bird, and the incapacity to fly may have positive survival value. It is quite possible that not being able to fly keeps you down at ground level, and not up in the air where any vagrant storm could pick you up and whisk you away. A number of birds have "chosen" that direction. Others have "traded" the power of flight for specializations in other and even more remarkable directions. The penguin now flies under the water, and the ostrich has become geared to great size, great strength, long legs, and running power.

The most noteworthy of the birds which lost the power of flight were the dodos and solitaires of Mauritius, Rodriguez, and Reunion Islands of the Mascarene group in the Indian Ocean. These hyperspecialized and thus defenceless creatures were related to the pigeons. Pigeons are among the world's greatest fliers, but the island birds went their evolutionary way in the direction of sheer size, while their wings atrophied. They were no match for the European sailors who inevitably arrived on their islands, and theirs were the earliest extinctions for which dates are available.

Mauritius and Reunion were "discovered" by European sailors in the first decade of the sixteenth century, at about the time of Balboa's Panamanian success. The lamented dodo of Mauritius had become extinct by 1681. The Reunion solitaire lasted until 1746, and the solitaire of Rodriguez until 1791; but the first appearance of man on their islands signed and sealed their sentence. It has been the same world-wide. Since Columbus chanced upon the West Indies, some forty-six forms of birds and mammals have become extinct in those islands. At least forty-one more are on the critical list today. All of them would still be thriving, had there been no navigation. The age of exploration ranks with the period of Pleistocene overkill and the current technological orgy as a landmark in the impact of man on wild nature.

When we recognize the immense self-confidence of Renaissance man, his new navigational aids, gadgets such as the fourteenth-century Chinese import, gunpowder, and his growing efficiency in the applied sciences, it is not surprising that when his self-esteem was reinforced by religious orthodoxy, he set about spreading that orthodoxy. If necessary, he would spread it by force.

Whether mercantilism or religious conquest was foremost in Europeans' motives, it is difficult to say. The upshot was, however, especially in the Catholic seafaring countries, a zeal for subdual which has only been equalled since in the invasion of the moon. Might was inevitably right.

There is colossal arrogance in the phrase "Age of Discovery." In the sixteenth and the seventeenth centuries, Europeans "discovered" not only new lands, new faunas, new animals and plants, but whole civilizations. The assumption was that these phenomena had been placed there, by almighty providence, for the discovery, edification and profit of Christian Europeans, and of the institutions they represented.

At this distance in time, the events which followed the 1519 landing of Hernan Cortes at the beautiful and tranquil island of Cozumel seem incredible. Yet it is possible to find a modern counterpart. Rolf Edberg reminds us that "In Montezuma's capital city, Cortes kills as many people in the name of His Highest Catholic Majesty as atomic bombs were later to sweep away at Hiroshima and Nagasaki." But Cortes did not have nuclear fission. No one knows how many Indian people died in the course of the Spanish take-over, because it is not certain how many there were beforehand. There have been recent estimates of perhaps 40 million for Latin America before the conquest. For Mexico, it is reckoned that the population at the time of Cortes' arrival was about 11 million. By the end of the century, it had dropped to 2.5 million. Another estimate has put the population of central Mexico at 25 million pre-Cortes, with a drop to a little over one million by the end of the century. Whatever the absolute figures, the depopulation was unprecedented, save perhaps for the effects of the height of the Black Death in Europe.

Butchery on this scale would seem to be beyond the physical capacity of even the most vigorous will. The invaders had something else working for them, what J.H. Elliott calls "that grimmest of all *conquistadores*" – smallpox.

An introduced disease, like any introduced organism, can play havoc with people – or other animals – which have not been conditioned to living with it. Just as introduced rats, goats, swine and other satellite animals of man disposed of the unique faunas of so many oceanic islands, smallpox (and measles) disposed of vulnerable people who had not lived with the diseases long enough to have developed a natural immunity to them. There is also evidence that following the ravages of smallpox in Mexico "social disorganization seems to have paralyzed recovery" (Ohlin). As in all natural situations, a homeostasis or equilibrium exists in human and other animal societies. It has been arrived at over many thousands of years of slow development. Sudden and widespread change, introduced over a short period of time, can destroy a society and reduce its chances of recovery to the point of no return.

The Spanish Christians happened to be Catholics, but the Catholics did not have a corner on vengeful and fiery retribution for the heathen. Protestants were equally savage, and, with the Catholics of the sixteenth century, believed that the proper penalty for heresy was burning. Protestant or Catholic, the Christian of the time viewed the non-Christian "pagan" as a kind of savage creature, a species of "animal" whose welfare was thus unimportant. While the heathen was outside the persuasion of Christianity, he did not exist as a human being. And if he was not a human being – he was nothing. It follows that the Christians' attitudes toward animals and wild nature were accordingly arrogant or indifferent.

W.H. Prescott revealed an interesting difference between the two Christian beliefs. He noted that the Catholics felt a duty to fight the battle of Christ in order to rescue pagan souls for eternity. The Protestants, perhaps even more egocentric in their view, considered that "Providence never designed that hordes of wandering savages should hold a territory far more than necessary for their own maintenance, to the exclusion of civilized man." Today, civilized men of all faiths hold true to that view, as any Eskimo can attest.

Trevelyan, in speaking of Drake, put it another way. Drake "wanted no inch of Spanish soil in the old world or the new. His objects were booty, trade, freedom to sail the seas and worship God aright, and ultimately to colonize empty lands where the Red Indian nomad would be the only person aggrieved." All of which, for both Spaniard and Englishman, reflects the colossal insensitivity of Christianity to the non-Christian world, whether it be a world of human or nonhuman beings.

The traditions of Western ethnic strains in relation to nature, wildlife, and conservation are intriguing. There is a pattern of distribution of the "nature ethic." Though the tradition is insufficiently developed anywhere, there are recognizable differences. The "best" of conservation awareness and practice is found in those parts of the world which have been influenced by Scandinavian, Anglo-Saxon, and German traditions – in general, the cultures of northern Europe. Also in general, the "worst" of conservation awareness and practice is conspicuous in those places largely dominated by French, Spanish, Portuguese and Italian traditions and cultures.

There are exceptions but, on a broad canvas, the breakdown is clear between northern Europe and the Romance traditions. There is a parallel contemporary difference between Protestant and Catholic traditions. It is no accident that the modern bullfight – surely one of the more horrendous manifestations of human attitudes toward nonhuman beings – continues in those countries where it does.

The appeal of the bull-ring was expressed by Ernest Hemingway, an *aficionado* for his own reasons: "A growing ecstasy of ordered, formal, passionate, increasing disregard for death. . . . It is impossible to believe the emotional and spiritual intensity and pure, classic beauty that can be produced by a man, an animal, and a piece of scarlet serge draped over a stick" (*Death in the Afternoon*). There are interesting insights in that passage – spirituality and, of course, the separation of man and animal. Christian spirituality is often as bound up with conquest as it is with the man/nature dichotomy.

It is difficult to detect any significant difference between the missionary zeal of the sixteenth- and seventeenth-century conquerors and that of their twentieth-century counterparts.

There is little discernible difference between the invasion of Central and South America and the invasion of the moon, or of the arctic tundra. All represent the advancement of a faith — and it is in man's most radical being to stop at nothing in this pursuit.

The GREAT EGO TRIP

Man is the introspecting animal. The possession of self-awareness and self-consciousness is included among definitive descriptions of our species. If there is one thing that distinguishes man from other animals, it is his mind. In it dwell thought and feeling, consciousness and volition; as St. Thomas Aquinas saw it, "those faculties of the soul which in their operation dispense entirely with matter." Wordsworth proclaimed the mind of man a thousand times more beautiful than Earth itself.

Of all his firmament of attributes, of all the arsenal of his skills and potentialities, none is so important to man as his Reason, which a generation after the death of Christ, Epictetus held "shapes and regulates all other things." Somewhat later, George Bernard Shaw was to observe, "The man who listens to Reason is lost; Reason enslaves all whose minds are not strong enough to master her." Reason is heady, and at least occasionally intoxicating.

The worst of Aristotle – his man-dominated ladder of increasing perfection – was rationalized with traditional

Judaeo-Christian dogma by St. Thomas Aquinas. After Aquinas, the anthropocentric intellectual "construct" was elaborated upon significantly by Francis Bacon and René Descartes.

It was Bacon's creed that "knowledge is power." Knowledge is the "perpetrator of man's empire over the universe." Bacon is the patron saint of the technocrat. His unshakable belief was that "the true and lawful end of the sciences is that human life be enriched by new discoveries and powers." Always powers.

Bacon's preoccupation was with causes. He observed an effect, then worked back toward the cause or causes. He would then experiment to see whether his discovered cause produced the observed effect. It was an ultra-mechanical approach which he would have applied, had he lived, to *all* things – not merely the sciences. Thus explained in the second (1620) of his anticipated six books, Bacon's method was to be applied "to all the phenomena of the universe." It is still the human assumption that Baconian methodology, through sheer Reason, can be applied to all the phenomena of the universe. With Bacon, the Great Ego Trip was under way.

Bacon was very much a theist, which fitted nicely with his highly evolved homocentricity: "They that deny a God destroy man's nobility, for certainly man is of kin to the beasts by his body; and, if he be not kin to God by his spirit, he is a base and ignoble creature." The separation of God/man/nature was fundamental. In *The Advancement of Learning*, Bacon observes that "when the mind goes deeper, and sees the dependence of causes and works of Providence, it will easily perceive, according to the mythology of the poets, that the upper link of Nature's chain is fastened to Jupiter's throne."

The Great Chain of Being, as Arthur O. Lovejoy characterized it, is a notion that is pre-Aristotelian. Like all the other conceptual pictures of life and the universe, it proceeds in linear progression from beginning to end. The end is sometimes man, sometimes God, but man is always given superiority over the rest of nature. The chain analogy is of course the source of the irritating and quite erroneous notion of the "missing link," which implies that there can be, and was, a definable stage between *Homo sapiens* and his predecessors. Far from being an orderly determined progress, evolution is a three-dimensional searching and probing process with events occurring not sequentially but simultaneously. The direction of evolution is toward diversity, not linearity. It is quite obviously satisfying to the collective human ego, however, to see a long and orderly succession of linkages terminating with ourselves, or with a god in the anthropomorphic image. The Great Chain of Being, like Aristotle's ladder and Teilhard's tree of life, is a comforting and reassuring picture.

Garrett Hardin describes the Great Chain of Being: "All living organisms represent stages in the idea of 'Being,' each intermediate form being a more advanced stage than some, and a less advanced stage than the rest. If we had before us all the organisms that have ever existed, we could arrange them, single file, in a continuous chain of increasing complexity." For the atheist, the last link is man; for the theist, God. Either way, man's superior position is conspicuous by comparison with those of lesser creatures.

It was Francis Bacon's belief that inquiries into scientific and religious matters should be kept separate. For him, philosophy has three objects: God, nature, and man. They should not be confused, because they are discrete. Later in the seventeenth century, science and religion were combined in such ways as John Ray's 1691 *The Wisdom of God Manifested in the Works of the Creation*. For Bacon, however, God had provided

two quite separate and different avenues of approach to revelation – the Bible, and the physical world.

Bacon made science "respectable," in the sense that
during the Middle Ages science had been mostly restricted to
alchemy at best, black magic at worst. Science had been
widely considered profane. Bacon sanctified science, which
now became the revelation of God's acts, and "the exaltation
of God's glory."

A generation younger than Francis Bacon was René
Descartes, the greatest of all apostles of Reason. So confident
was Descartes in the reasoning capacity of the human mind
that he felt it would be possible to discard all books and then
to reason the universe into understandable terms. He
rejected most scholastic tradition and all theological dogma.
But it seems that he threw out the baby with the bath-water
when he also disposed of all the joys of intuition. Bacon had
been confident of man's ability to Reason and thus to "end in
certainties," but Descartes was downright vainglorious. For
him, knowledge would allow men to become "the masters
and possessors of Nature."

If Bacon was the first of the modern scientists, Descartes
was the first of the modern philosophers. His *Discours de la
Méthode*, published in 1637, is a detailed account of his long
and passionate affair with his own mind. He refused to believe
anything in books unless it was supported by what he called
"incontrovertible and absolute" proof. Descartes soon discovered that the only thing that could stand his test of doubt was
the fact of his own existence. Thus, *"Cogito: ergo sum."*

For Descartes, it was vital to separate mind and matter.
All things fall into one of these two categories, and neither
can influence or benefit from the other. By separating mind
and matter, Descartes was also able, as at one stroke, to separate religion and science, and to allow scientific inquiry to

proceed without the limiting inhibitions of dogma. There was no doubt whatever in the mind of Descartes that all things can be discovered and known by the Rational Soul. As he says, ". . . all things, to the knowledge of which man is competent, are mutually connected in the same way, and there is nothing so far removed from us as to be beyond our reach, or so hidden that we cannot discover it. . . ." Descartes thus provided another rocket stage for the Great Ego Trip. We were ready for Descartes, and we embraced him. He was properly sanctimonious, he was properly scientific. He had further sanctified uniquely human Reason, and thus man.

At the same time, in another part of France, there was Blaise Pascal, who believed that "to think well is the principle of morality." Pascal – brilliant, religious, mystical, and "sublimely" articulate – was possessed of a humility which was not common in the Age of Reason. "The eternal silence of these infinite spaces (the heavens) terrifies me." He also acknowledged that "the whole visible world is only an imperceptible atom in the ample bosom of nature." Such remarks prompted many of his contemporaries and successors to brand him as a pessimist! As a man, he could not of course move from his position of uniqueness, but that limitation was not unique to him, nor to his time.

And then there was Thomas Hobbes – materialist, mechanist, political theorist. Like Bacon, he was persuaded of the immediate and future practical values of science and knowledge. Like all the others, he separated man and nature, and treated these in turn separately from supernatural inquiries. With Alexander Pope, he believed that "the proper study of mankind is man," and got on with his rationalist materialism somewhat to the dismay and resentment of the religious. For Hobbes, Reason is the covenant that binds together human societies, for without Reason man is "poor, nasty, and brutish," motivated entirely by appetite and desire, with his natural state being constant strife, enmity and war. Only Reason is his salvation.

Despite Hobbes and the "killer ape" theory and the "fall of man" notion, our primate roots are not in strife, enmity and war. Rather, they are in tranquillity and inoffensiveness. Hobbes' was a sterling defence of the advancement of science, but perforce it carried forward with it the tacit assumption of the uniqueness of man and Reason in the universe, and thus the "superiority" of man over nature. Nature was in no way permitted to intrude upon his thoughts except as the raw material for the advancement of human goals.

No one doubted that man's essential nature was his rational soul. Scientists and philosophers, whether religious or not, had Reason as their common bond. Newton, a strangely religious man (and a practicing alchemist) was, like Pascal, possessed of an aberrant humility: "I do not know what I may appear to the world, but to myself I seem to have been only a boy playing on the sea-shore, and diverting myself in now and then finding a smoother pebble or a prettier shell than ordinary, whilst the great ocean of truth lay all undiscovered before me."

One can only surmise on the extent of the anguish which may have been in the mind of a man who had such conflicting theological and scientific views. It is known, however, that it took the greatest tact on the part of his great friend Halley to persuade Newton to publish his *Principia*. Much of it would probably have been suppressed by Newton; he feared a religious controversy. A similar struggle with conscience was to be the fate of Charles Darwin, brooding in his study for so many years during the writing of *The Origin of Species*, never becoming entirely emancipated from the religious tradition of his childhood. Darwin felt as though he was "committing murder" by even entertaining the thought of a mechanism for evolution, much less publishing it.

During the golden period of evolving science, the thread which ran throughout was Baconian causation. Newton did not know how to handle the ultimate causes or origins of things. He had postulated a cosmic mechanism that operated according to universal laws. It did not require constant force or power, for it was self-propelling. God was needed only to start it. But just in case some constant power source *were* needed, Newton, in Bernal's words, "left a loophole for divine intervention to maintain the stability of the system."

Newton, a Platonist, believed in a Supreme Being, and thus had to allow that Being a role in his cosmology. (It was in this way, it is said, that Platonism entered the mainstream of our Western thought – by way of Newtonian physics.) Newton, who was sufficiently honest to admit his ignorance about original causes, nonetheless believed in the ultimate *necessity* of original causes, and had to leave things to the will of God at the beginning of Creation. Mechanistic scientists have always been embarrassed by this, but it is difficult to fully understand their embarrassment. If there *were* beginnings of the cosmos, there is no evidence to refute that they were the will of God. In the framework of the cosmic instant which will be the career of the species man, original causes seem scarcely relevant.

Reason presupposes causes, and causes presuppose necessities. The occasional dissenter, such as David Hume, has been largely ignored. Hume denies causes and effects: "Necessity is something that exists in the mind, not in objects." He also denies any necessity for "self." He feels that the pursuit of Reason has succeeded only in bringing us "into a frame of mind where the solid fabric of the world dissolves like a dream before our eyes, or passes into a kaleidoscopic unreality of change."

Human Reason has had a long history, and the Great Ego Trip is far from over. We have made it to the moon, and we have sent a space vehicle to circle Mars, and photograph it. Such notable accomplishments add immense impetus to the self-glorification of the reflecting primate.

Primates as a group are "generalized" creatures. They can eat almost anything within broad limits; they can live virtually anywhere. Thus, they have been so successful over tens of millions of years. Man has parted company from the other primates in only one major way – the development of his brain. In this sense, man is no longer generalized but highly specialized.

There are other animal species which have large brains, notably whales. We do not understand the brain of a whale, but experiments with dolphins and other smaller whales reveal that there is a good deal more cerebration going on than we anticipated. Dolphins and others can be "taught" – or encouraged – to do remarkable things, and when one watches them for a time there is no doubt whatever that they are capable not only of reasoning but also of "feeling."

It is strange, given the size and complexity of their brains, that the largest whales are so pathetically easy to kill. The much smaller and less convoluted brain of a crocodile or alligator allows the animal to quickly learn about people, jacklights and guns. Ducks and geese know where shooters lurk. Why is it then possible for a whaling fleet to wipe out whole pods of whales, one after another? These are thought to be the most intelligent nonhuman animals on Earth. Why do they not profit by observing the fate of their kind?

The largest whales are at the top of the ocean food-chain; they are the great grazers of the oceans. They have few, if any, important natural enemies. Occasionally, killer whales and some of the biggest sharks may take their newborn young, but this would not seem to be significant in terms of total numbers. Perhaps whale brains, because of their possessors' almost absolute security, do not need to conceive of fear. Perhaps, in its innocence, a whale can only conceive of love. There are many fully documented observations, both in the wild and in captivity, of whales such as dolphins which have demonstrated, not only to their own species but also to *other* species, that trait we call kindness. They have helped other kinds of whales, and they have also helped people. Perhaps the brain of the very largest whale is *over*specialized toward kindness and altruism, in that it cannot comprehend the notions of fear, or mistrust − or evil. Overspecialization, either behavioral or physical, can carry the seeds of eventual extinction. An animal with a peculiar and overdeveloped speciality, such as complete dependence on eucalyptus leaves, is surely doomed if eucalyptus leaves for some reason are not available. An animal with any highly developed speciality is doomed if the environment should suddenly change − perhaps ever so slightly − making that speciality no longer appropriate, or even advantageous. It is quite possible that the human brain may be such an overspecialization. Reason may be such a liability.

A strong case may be made for Reason as an over-specialization. Reason gave us the technology which is killing other species of animals and plants, killing us, and killing planet Earth. Reason allowed us to rationalize the divine mission to subdue. Unlike the blue whale, Reason gave us the capacity to conceive of evil. Reason gave us the concept of species hierarchies and species dominance, a perversion of the natural social dominance of other primates.

In the sixteenth century, Montaigne reflected that science without conscience is but death of the soul. Montaigne was no technocrat: "Let us a little permit Nature to take her own way; she better understands her own affairs than we." Montaigne, like Hume and Pascal, represents the positive side of Reason – the warmth of wisdom.

As actions have their reactions, so emotions have their counter-balancing opposites: love/hate, hope/despair, joy/anger. Neither side of either pair can exist without the other. Such are the specializations of the human brain which have evolved side by side, simultaneously and con-currently, through time. They have been there as long as our brains have been human. The moment we could love we could also hate, and the moment we knew despair we were expressing the capacity for hope.

The evolution of Reason must perforce have involved both sides of the old combinations – positive and negative. We could not have had Renaissance cathedrals without the slaughter of Aztec and Inca innocents; both were in celebra-tion of the European church. We could not have had the advancement of science without the intellectual arrogance of the Age of Reason. These are the "trade-offs" which occurred along the ancient path of evolving culture. They are compara-ble to the inevitable trade-off the seal made, for example, when for success in the water it was compelled to forsake the ability to walk.

Reason is rather like a "technological fix." If it gets us into trouble, we attempt to apply just a little more of the same in order to extricate ourselves. Thus, frequently, we compound our problems. It is quite possible that this is the evolutionary future of our personal overspecialization. Like a peacock's tail, it will continue to grow, whether we like it or not (and, like the peacock, we are terribly proud of it), and it will reach a point at which it will no longer have survival value. From then onward, extinction is swift. Perhaps, with mindless and directionless technology, we have already passed the point of return – the achievement of humility.

Garrett Hardin takes a more optimistic view, arguing that intelligence is adaptive and thus unspecialized. "It is not like an instinct, finely adapted to an immediate solution to one problem. It is more a sort of skeleton key, capable of opening a wide variety of doors. But with it man may have opened one door too many and thus have led the way to his own extinction. We don't know. But if he does not kill himself, it will be because his godlike intelligence enables him to find but one more door."

The extinction of man is as certain as his existence. Extinction is certain for all species, faunas, planets and stars. Individual species, like individual beings, do not *matter* in the larger system of things. No individual or species will be selected for immortality. We will be mourned by no one. In Darwin's words – the world will go cycling on.

THIRTEEN

The FLUTES of ARCADY

One cannot be certain whether man is the only animal capable of nostalgia, but having in mind his well-developed capacity for self-deception, one might suspect that he is. This failing has been illustrated in religions and schools of thought from Aristotle to Teilhard. The latter, with only the best and loftiest of motives, was incapable of avoiding the ultimate self-deception — the position of man on the "tree of life."

R.D. Laing says, "Human beings seem to have an almost unlimited capacity to deceive themselves, and to deceive themselves into taking their own lies for truth." We do this not only as individuals, but also as groups, as societies, as cultures, and as a species. Man is the self-duping animal; in this respect at least, man is singular. This is especially true of nostalgia; things past which linger so glowingly in our remembrance rarely if ever existed — at least not in the forms in which we remember them.

Romanticism is nostalgia for things which never were. It is the ultimate in self-deception. We are all guilty of it, and at one time or another we all *know* we are guilty of it, but we persist in our romanticism. And, of course, we cannot face the possibility that there may never have been anything there in the first place. (That would be like denying that there is any Reason for things.) As Laing says, "Take anything, and imagine its absence." That is extremely difficult to do – if not impossible – because it is frightening. It is hard on our security. Laing adds the devastating thought, "'There's nothing to be afraid of.' The ultimate reassurance, and the ultimate terror." *Nothing?* Just as we are "afraid to approach the fathomless and bottomless groundlessness of everything," we cannot face the fact that there never was an Arcady.

One knows there was a place called Arcadia. It was not, however, Camelot. Arcadia was a plateau on the Peloponnesus which was cut off by chains of mountains from the rest of what is now Greece. Thus, its people did not mix much with those of the other countries around it, and did not acquire elements of their cultures. It is said that the people and culture of Arcadia "stagnated." They remained hunters and pastoralists, while their more advanced neighbours in adjacent states, such as Sparta, were busy making wars.

At some point, perhaps even before Vergil, Arcadia became idealized as the never-never-land of total human happiness – even beatitude. Idyllic conditions were imputed to it, and the carefree shepherds and shepherdesses became the envy, the delight, and the concreted nostalgic remembrance of a large part of the Western world. The likes of Cowper, Tennyson – even Goethe – recalled Arcadia and its joyous, childlike innocence.

There have, however, always been heretics among us – people who knew that *You Can't Go Home Again*; but most of us have preferred to bask in the remembered joys of sunny, bucolic Arcady. Like fantasying of any kind, romanticizing the past is inconsequential and relatively harmless. It becomes important and potentially destructive when, through romanticism, we impute to the nonhuman world uniquely human emotions, characteristics, and attitudes. Disney romanticizes nature. It is a very fine line between romanticism and anthropomorphism.

The poet who sees benignity and benevolence in nature – or harshness and cruelty – is denying the utter indifference of nature to human existence and human values. Natural objects and nonhuman animals know neither nastiness nor saintliness. (I have long suspected that the Romantic poet, rejoicing in the lily, is somewhat narcissistically rejoicing in his sensitivity to the "lower" forms of life.) The Romantic does not appreciate Alexander Pope's "Whatever is, is right." All must be seen in human terms, and all must be possessed of human values. Otherwise, it is not real.

Man is obsessively preoccupied with rank. Man is the ranking and grading animal. Modest men may posit angels and God; immodest men may see the final link in the Great Chain of Being or "inflorescence" as man. Either way, *verticality* is inherent and essential; man is man, and after man there are all the "lower" species. All things must be classified, and all must be labelled. All must be given their appropriate position in file, whether by God or by man, in an immutable hierarchy. We have never learned to think in more than two dimensions.

Perhaps our preoccupation with hierarchical structures is a demonstration of Alfred Adler's concept of our constant and life-long battle with inferiority. Robert Ardrey relates this to the struggle for dominance, the juggling and jockeying for position and status. Perhaps Adler's phenomenon also encourages us to picture ladders, chains, and ascending orders of perfection, from Omega to Alpha.

We know that other primate societies, such as those of baboons and chimpanzees, have dominance hierarchies. Such structures are necessary for societal survival. There is no difficulty in accepting that hierarchical thinking is an integral part of our primate heritage, and that nothing in our makeup is more natural than to attempt to dominate other members of our society. But man extends this drive not only to individuals of his own species, but to *other species*, a phenomenon unheard of in nature. If we are to acknowledge, as we must, that there are Alpha individuals, then it seems logical to our illogical minds that there must be Alpha *species* also. Human thought is riddled with this type of syllogistic reasoning. Adler's picture (and a perfectly accurate one) of the *intra*specific individual power struggle becomes Reasonable man's picture of *inter*specific dominance hierarchies, which is a false and unnatural one.

In eighteenth-century Europe the whole concept of "Nature" underwent surprising evolutionary development. Setting aside for the moment the self-centredness of the human position on the Great Chain of Being, the notion of a chain had promoted a new idea of unity in natural systems. All things were tied together in a "master plan," as though it were one great piece of machinery, operated by divine benevolence.

Thus we had the "divinization of Nature," as expressed in the view of the third Earl of Shaftesbury, who, like Newton, was a deistic Platonist out of Cambridge. "'True religion' should be based on 'Nature' rather than on Revelation . . . also, orthodoxy is the enemy of 'true' ('natural') religion, because it invites us to base our faith, not on the beautiful and harmonious Order of Things – the best and only genuine external evidence – but on miracles, that is, on infractions of that Order" (Willey). This, of course, was a bit much for the church to swallow, what with its traditional insistence upon revelation, but it was just going to have to manage. "Nature" was on the upswing.

"Nature" made it possible for a man to be religious whether he believed in God or not. The godly man saw nature as God's handiwork, and believed that God, "instead of expressing his will directly, was bringing it to pass through the realm of nature and natural law" (Soule). Those who had dropped religion were able to substitute the entire world of living nature. Those who had become disappointed or frustrated with man and his works were also able to substitute nature as the recipient of their loyalties. There was never any question, however, about who was the sole beneficiary of the grand design – Enlightened Man. Neither atheist nor churchman had any doubt about that.

There appeared, however, a growing body of people who were disappointed and frustrated by man, and angered at the accelerating expansion in human self-esteem. The widespread acceptance of the new religion of Nature, especially as Nature was revealed and understood through Reason, made this period an ideal climate for the emergence of a group of satirists, who assailed eighteenth-century civilized European man for his overweening "pride." The smile of Reason has spawned the vice of pride, and Pope was vigorous in his reactions:

> In pride, in reas'ning pride, our error lies;
> All quit their sphere and rush into the skies!
> Pride still is aiming at the bless'd abodes,
> Men would be angels, angels would be gods.

This was a reference to a certain amount of jostling for position which Pope had noted at the top end of the Great Chain of Being. Pride, the product of misused Reason, is throwing a spanner into the perfect machine.

All nature is but art unknown to thee.
All chance, direction which thou canst not see;
All discord, harmony not understood;
All partial evil, universal good;
And, spite of pride, in erring reason's spite,
One truth is clear, Whatever is, is right.

No criticism of the time was so savage as that delivered by Dublin's Jonathan Swift, whose moral indignation and sincerity of purpose drove him to attack not only politicians and prevailing social customs, but also the entire human species. For Swift, man has betrayed his own Reason, and through sheer stupidity and vanity has distorted his own values and even his own judgment. Swift reverses the traditional hierarchy and places the filthy, degenerate Yahoo at the bottom of the ladder. As Lovejoy describes it, "The most detestable and irrational of beings (man) crowns his fatuity by imagining himself the aim and climax of the whole creation." Two centuries later, man is still regarded as the aim and climax of the whole creation.

Anyone who has spent the greater part of a lifetime enjoying and attempting to understand and preserve wild nature will have had the experience of witnessing his own species drift lower and lower on his personal scale of perfection. All the magnificence and nobility of our creativity cannot begin to compensate me for what my species has cost me. Shakespeare cannot compensate me for toxic pesticides, Bach cannot compensate me for Hiroshima, nor Michelangelo for the blue whale. Jesus Christ cannot compensate me for the brutal imposition of human power over nonhuman nature. Yet, the total destruction of blue Earth may well precede any diminishment in human pride.

Misanthropy is probably the inevitable (or at least occasional) companion of the man who values unspoiled nature, natural systems, and their individual components. Byron could say, "I love not man the less, but Nature more," and could steal moments to "mingle with the Universe." He was aware that "Man marks the earth with ruin."

There was also a great wave of nostalgia. But it was not nostalgia for any early time that had actually existed; it was nostalgia for the word-picture of Arcadia. It was Romantic. In John Ruskin's words, the nostalgia was a "pathetic fallacy," for the new movement saw in Arcadia, and in nature, things that were never there. The Romantic imputes to nature emotions that are peculiar to the Romantic, not to nature. He is recasting nature in his image; he is anthropomorphizing it.

Feelings such as pleasure and grief are not unique to man. Nonhuman animals show affection, and they show sadness; but they do *not* attribute the capacity for these emotions to other animals, nor do they read mirror-pictures of themselves in the behaviour of other animals. That is a solely human characteristic, so far as we know. Animals do not zoomorphize, probably because they cannot conceive of interspecific dominance and thus have no need to recast the rest of the world in their own image.

Nostalgia generally involves a sweet kind of sadness, and there was plenty of melancholy in the Romantic movement. Jean-Jacques Rousseau and his followers were thoroughly disenchanted with the state to which men had brought eighteenth-century European society, and they bewailed the fall from grace that was represented by man's departure from a "state of nature."

Rousseau had great difficulty in defining a "state of nature." Certainly it could not be the "brutish" situation described by

Hobbes. That would have been unthinkable to the European mind of the time. But civilized man, in Rousseau's view, was even worse than Hobbes' beastly "primitive," which was little more than a Yahoo. So he was forced to make a compromise. As Leo Marx says, "Unable finally to endorse either the savage or the civilized model, Rousseau was compelled to endorse the view that mankind must depart from the state of nature – but not too far." Marx goes on, "Curiously enough, Rousseau thought that mankind had passed through the ideal state during the pastoral phase of cultural evolution by which he meant a pastoral situation in a literal, anthropological sense: a sense of herdsmen." Rousseau believed not in original sin but original good. Man the pastoralist was seen as man in nature, and that is where he should have stayed.

Rousseau's arrival at this position was, ironically, a prime example of the use of Reason in rationalizing a highly complicated point of view. Ardrey has been hard on Rousseau for this. "The organizing principle of Rousseau's life was his unshakable belief in the original goodness of man, including his own. That it led him into the most towering hypocrisies . . . is of no shaking importance; such hypocrisies must follow from such an assumption. More significant are the disillusionment, the pessimism, and the paranoia that such a belief in human nature must induce."

Surely, however, the assumption of original good is no worse than the assumption of original evil, or of any other starting point. It is now in our tradition, even as we move toward the final quarter of the twentieth century, that there have always been beginnings, and causes, and directions. It is difficult to single out the eighteenth century for castigation on the basis of such assumptions. Hobbes was just as wrong as Rousseau, and the twentieth century is just as wrong as the eighteenth.

Romanticism is a great deal easier to accept than Reason. Romanticism at least sees some value, even if it is only a human-centred value, in the nonhuman world, and admires things which are not the products of Reason. The longing for Arcadia is pathetic, but the worship of human intellect to the exclusion of sensibility is preposterous and potentially suicidal. Daniel McKinley has suggested the "possibility of a core of sense in what may have seemed just another fad," because romanticism may have "generated some salvageable truths that previous centuries had failed to disclose. . . . Romanticism, in its reverence for the integrity of the exotic and its presuppositions of a divine fullness in kinds of things, helped lay the groundwork for modern ecology and natural history."

There is, however, a danger inherent in Romanticism. The nostalgic view of nature, with man set in it in the form of a gentle pastoralist, does just as much to separate man from his environment as does the cult of the cold intellect. Both are anthropocentric. Both see nature and man as separate entities, albeit in different relationships. The Romantic of Rousseau's school sees man using nature as a plaything or as a backdrop for the display of man's most admirable characteristics. Nature is the mirror for this self-admiration.

Irving Babbitt has made this point with conviction. "The Rousseauist does not in his 'communion' with nature adjust himself to anything. He is simply communing with his own mood. Rousseau chose appropriately a title for the comedy that was his first literary effort *Narcissus or the Lover of Himself*. The nature over which the Rousseauist is bent in such rapt contemplation plays the part of the pool in the legend of Narcissus. It renders back to him his own image. He sees in nature what he himself has put there. . . : Nature is dead, as Rousseau says, unless animated by the fires of love."

The Romantics were forever moralizing about the natural world; they were determined to find "good" in nature, and to show how contemporary man was "evil" by comparison. By extension, Romanticism in the "Nature" context becomes akin to religion. It comes very close to being institutionalized, complete with dogma. This seems to be the inexorable evolutionary trend of almost any body of thought, or even of feeling. All must be codified. Nature is virtuous; ergo, man is unvirtuous, because man has left nature. The separation between man and nature is categorically described, as it is in our organized religions. Like any other creed, Romanticism offers simultaneous threats and promises, giving with the one hand and taking away with the other.

Granting the best motivations to Rousseau's cause (with which any naturalist can only sympathize), it points up a fundamental characteristic of mankind — the compulsion to forever *evaluate* things, to *pass judgment*. This compulsion seems inescapable. Men of Reason pass judgments, as do Romantics. Scholars moralize, as do poets, escapists, misanthropists, conservationists and technocrats. We are driven by the urge to categorize things. And more often than not, we use the age-old scales of perfection. This is better than that, and that is worse than this. And we recommend cures for erring human nature. Finally, we evaluate all things only in human terms, and by human standards.

Even our language is constrained by the subjective judgments inherent in the words we use. Take that innocuous word "humane." The very best of dictionaries defines "humane" as civil, courteous, obliging, kind, benevolent; it evokes mercy and compassion. These are the properties inherent in the root word "human." On the other hand, we find no mention of the rest of the long list of uniquely human attributes so painstakingly spelled out by Hobbes, Swift, and even Ambrose Bierce. Where are uncivil, discourteous, selfish, cruel, malevolent; and where are merciless and pitiless?

We have amputated one-half of the human equation and expunged it from our language. It is no wonder we have such great difficulty in defining "human nature"; we no longer have the words for it, unless we look up "paradoxical" or "schizoid," or some such. The Romantics saw only one side of human nature, the misanthropists only one, the men of Reason only one. Has any man been able to see all sides at once, save the rare likes of a Shakespeare or a Pascal? Perhaps only the cubist painter is capable of that.

It is a melancholy fact that many of those who extol the virtues of nature are themselves seriously out of joint with their human society and surroundings. The aphorism "Nature is nobler than man" is just as fatuous as "What did nature ever do for me?" We have heard both points of view. I have heard the former from the occasional recluse who, because of a personality short-circuit, is estranged from the human community. He is afraid of people and he knows that a sparrow or a toad cannot hurt him. Like the pet-fancier (if Adler is right), he indulges his frustrated dominance drive in that nonhuman direction. Such a posture is pitiable.

The position "What did nature ever do for me?" is more than pitiable; it is destructive. Estrangement from nature is estrangement from oneself, because one *is* nature. One is also posterity. The publisher (or the chairman of the board) of a major U.S. popular magazine recently asked publicly, "What did posterity ever do for me?" The rhetorical question was raised in the context of some widely publicized environmental issue of the time. Such a statement is barely Neanderthal, but the business press is replete with them.

The Romantic movement was filled with affectation,

posturing, self-conscious sensitivity and sensuality, misanthropy and empty nostalgia. Yet it revealed other ways of
being human, of examining the phenomenon of man. The
flutes of Arcady were never there, but then, neither was
original sin. There was no "noble savage," there was no
Yahoo. There was only introspecting, self-deceiving man,
frantically searching for homocentric meanings, purposes,
and designs, when all the time there were none. The enormity of that simple truth has yet to dawn on us.

FOURTEEN

MISSION ACCOMPLISHED

From the earliest days of North American colonialism, the mission appeared multi-faceted, but its essential purpose was profit. Only the anticipation of profit could have prompted the backers of the *Mayflower* voyage in 1620 to underwrite the transportation of 102 Protestant dissidents to the new land. (It is ironical that the Puritans, so vexed by religious intolerance in England, were to prove less than tolerant once in the New World.)

The spirit of John Calvin met and conquered the immense North American wilderness out of sheer perseverance – and the sense of mission. Calvin was an indomitable man, the model of tenacity. He was also possessed of rare imagination. But even Calvin, at his death in 1564, could not have foreseen the colossal level of achievement which his model of dogged industry would inspire – the subjugation of an entire continent which had been revealed to Europeans only seventeen years before his birth.

There is little doubt that the Pilgrim Fathers and their financial backers had carefully read tracts such as Alexander

Whitaker's *Good Newes from Virginia*, published in London in 1613. Who – most especially a mercantilist – could resist such intoxicating reports as: "Wherefore, since God hath filled the elements of earth, aire, and waters with his creatures, good for our food and nourishment, let not the feare of staruing hereafter, or of any great want, dishearten your valiant minds from comming to a place of so great plentie: if the Countrye were ours, and meanes for the taking of them (which shortly I hope shall bee brought to passe), then all these should be ours: we haue them now, but we are fain to fight for them, then should we haue them without that trouble."

The instigators of settlement in the new continent used every device, including artists and the church, to further their cause. As Louis B. Wright has put it, "A share of stock bought in a colonial company would be a stroke for both God and King, for the advocates of settlement in North America solemnly asserted their high purpose of saving heathen souls and establishing outposts against the extension of the Spanish empire." In 1622 there occurred an Indian massacre which laid waste a number of frontier farms, and the Virginia Company "persuaded no less a person than Dr. John Donne, the poet and dean of St. Paul's, to preach and publish a sermon declaring, somewhat metaphysically, that support of the colony would advance both England and the Kingdom of God. For this useful message, Donne received stock in the Virginia Company and for a time nourished the hope of becoming secretary of the corporation." Donne also contributed a short poem to accompany the publication of John Smith's account of New England, dedicated "To his friend Captaine Ion Smith, and his Worke."

The settlers were not long in noticing the stupendous abundance of American wildlife. There have been many published descriptions of the flights of passenger pigeons, but one of the earliest is in Wood's 1634 *New-England's Prospect*: "The Pigeon of that Countrey is something different from our Dove-house Pigeons in *England*, being more like Turtles, of the same colour; but they haue long tayles like a Magpie. And they seeme not so bigge, because they carry not so many feathers on their backes as our *English* Doves, yet are they as bigge in body. These Birds come into the Countrey, to goe to the North parts in the beginning of our Spring, at which time (if I may be counted worthy, to be beleeved in a thing that is not so strange as true) I have seene them fly as if the Ayerie regiment has beene Pigeons; seeing neyther beginning nor ending, length, or breadth of these Millions of Millions. The shouting of people, the ratling of Gunnes, and peltime of small shotte could not drive them out of their course, but so they continued for foure or five houres together. . . ."

The Reverend Francis Higginson had already mentioned the great "Flockes of Pidgeons" in his 1630 *New-Englands Plantation. Or, a Short and True Description of the Commodities and Discommodities of that Countrey*, but he clearly found the wild turkeys more to his taste. "Here are likewise abound-ance of Turkies often killed in the Woods, farre greater than our English Turkies, and exceeding fat, sweet and fleshy, for here they haue aboundance of feeding all the yeere long, as Strawberries, in Summer all places are full of them, and all manner of Berries and Fruits." The English turkeys to which he referred were descendants of the same American species, brought to Europe from Mexico by returning *conquistadores* and introduced to Britain about 1540. Higginson also rhap-sodized about the Massachusetts air, which must have been a godsend after the sulphur dioxide coal fumes of grimy London: ". . . a sup of *New-Englands* Aire is better than a whole draft of old *Englands* ale . . ."

Whether turkeys or pigeons, deer or bear, wildlife species

were there for the consumption of the God-fearing settlers. John Josselyn, in *New-Englands Rarities*, which was published in London in 1672, provided a long list of American birds and mammals, and the uses for which they had been provided. One of the most interesting of these concerns *The Wobble*. "The *Wobble*, an ill-shaped Fowl, having no long Feathers in their Pinions, which is the reason they cannot fly, not much unlike the *Penguin*; they are in the Spring very fat, or rather oyly, but pull'd and garbidg'd, and laid to the Fire to roast, they yield not one drop." The Wobble is the now-extinct great auk, which flourished in large numbers on its remote North Atlantic islands until Europeans arrived in sailing vessels. The flightless birds persisted until the 1840s, when the last individuals disappeared forever beneath the clubs of fishermen.

Josselyn had a medicinal application for great auks. "Our way (for they are very soveraign for *Aches*) is to make Mummy of them, that is, to salt them well, and dry them in an earthen pot well glazed in an Oven; or else (which is the better way) to burn them under ground for a day or two, then quarter them and stew them in a Tin Stew-pan with a very little water." He even had a use for the sharp, curved bills of ospreys: "Their Beaks excell for the Tooth-ach, picking the Gums therewith till they bleed."

Not all, however, was strawberries and roasted fowl. The land was hard. In the seventeenth century, the bleak Calvinist clergyman Michael Wigglesworth, author of *Day of Doom*, was moved to identify "God's Controversy with New England." Convinced that New England had been singled out for divine displeasure, the glum Wigglesworth reached back to Deuteronomy for his description of the land as a "waste and howling wilderness." This was to be the consensus for some time to come. But it was understandable. Peter Matthiessen has the pulse of the matter: "One must keep in mind that the glad reports of this period were motivated in part by the colonists' desire to share with others an existence made miserable as much by their own religious strictures as by hardship!"

Tight-lipped and unsmiling, the Puritans went about their business of conquest. The process was less spectacular than the blood-lusty deeds of the *conquistadores*, but the Puritans had come to stay, not to pillage. They had come to master, and it took a special kind of people to master wild North America. They had become that special kind of people long before they left England, because, as George Grant says, "When one contemplates the conquest of nature by technology one must remember that that conquest had to include our own bodies. Calvinism provided the determined and organized men and women who could rule the mastered world. The punishment they inflicted on nonhuman nature they had first inflicted on themselves."

Frontier people always fear and mistrust the wilderness, and usually they have good reason to. A subsistence existence is a precarious one, and even a relatively minor accident — a sprained ankle or a touch of frostbite — can have disastrous consequences. Considering the level of technology at the time, one can understand the pioneers' dread of the wilderness. However, the emotional evolution from fear to hate is less comprehensible.

Roderick Nash has identified what he calls a "tradition of repugnance" for the American wilderness. This tradition persists on the frontiers of our time. It is not easy to determine its origins. Certainly the Judaeo-Christian tradition is paternalistic and condescending toward wild nature, as well as aggressive, whether one chooses either the "dominion" or the "stewardship" interpretation of the Bible. But it is not openly hostile. Hatred of the wilderness as expressed by the colonizing Protestants seems to have been their own special aberration, without valid biblical precedent. Of course it is easy to read almost anything into many passages of both Testaments, and it would appear that the settlers seized on such texts as they needed to justify the holy war on nature.

The pioneers came to regard wild nature as not only dangerous but malevolent. Fortified by an obligation to dominate in God's name, the mission to destroy became obsessive. Thomas Merton has remarked that the settlers came to hate the wilderness "as a *person*, an extension of the Evil One, the Enemy opposed to the spread of the Kingdom of God." The Christian duty was to dispatch the enemy with speed and efficiency. The subjugation of nature became a crusade, prosecuted with ferocity.

The crusade required all that was best in the Calvinist character – grit, stamina, stubbornness, belief – plus lack of imagination and sensitivity. It required mission-oriented indifference to consequences, and zeal fired and stoked by God's will. Once it creaked into motion, the accelerating juggernaut was unstoppable. All gave way before it – pine and oak forests, grasslands, wildlife.

As Charles Darwin knew, "Ignorance more frequently begets confidence than does knowledge." Though the pioneers were prepared to endure severe hardship, they never questioned that they would prevail. The momentum of Right would carry them through, if they practiced diligence, thrift, etc.

There was a strange irony in this, described by Nash as the fact that "success necessarily involved the destruction of the primitive setting that made the pioneer possible." With the wilderness went the pioneer, and with the pioneer went the archetypic Puritan. A new North American emerged, characterized by an odd ambivalence about the land. On the one hand, the new master of nature carried the lingering heritage of hostility and contempt for wilderness, yet on the other he had imported from Europe a romantic vision of pastoral nature no less strange than that of Rousseau himself. This was a man seriously divided, and he persists to this day.

A splendid and penetrating analysis of the American schizophrenia about the land has been contributed by Leo Marx in *The Machine in the Garden*. Like European Romanticism, Marx says, the American pastoral tradition has man very much outside nature, but observing nature and deriving pleasure from it. "What is attractive in pastoralism is the felicity represented by an image of a natural landscape, a terrain either undespoiled or, if cultivated, rural." This is at odds with the dreadful desolation described by many of the Puritans, but Marx provocatively suggests that the image of hideous wilderness was exactly what the Puritans *wanted*. "Colonies established in the desert (as opposed to the gentle garden) require aggressive, intellectual, controlled, and well-disciplined people." The ugly picture of the wilderness was a challenge to the exercise of will and power, and it was a challenge that was eagerly met.

Throughout North American history, there has never been any reason for the existence of wildlife species save to serve man — whether for food, clothing, amusement, or profit. If a species did not serve human purposes, of what possible "good" could it be? If, as sometimes appeared to be the case, a species actually "competed" with human interests (too frequently, recreation), then clearly it must be eliminated. For some inexplicable reason, God, while creating doves, squirrels, and grouse, had also created coyotes, hawks, and rattlesnakes. This is what Jack Olsen calls "the Mother Goose table of animal values (wolves and bears are evil, bunny rabbits and chipmunks are good, etc.)."

Jack Miner, friend and patron of the Canada geese, called the great horned owl a "heartless cannibal," and made it his business to kill twelve to fifteen of the birds every winter. Divine wisdom had somehow slipped up in visiting such demoniac creatures on man's world of nature. But no matter. It was man's holy mission to clean things up.

The bison was obviously no cannibal, but he was there for human purposes; and anyway there were so many that they actually got in the way. Buffalo Bill Cody, who boasted of killing 4,820 bison in one twelve-month period, was the idol of rich and poor. He turned his personal legend into a travelling sideshow which earned him fame on both sides of the Atlantic. The Boone and Crockett tradition evolved into something close to sacrosanct, and the firearm cult is still very much alive today.

Killing for amusement is a deep-rooted human tradition. Early man was a hunter, and he became a good one. But when people began to settle in more permanent communities and undertook pastoralism and agriculture, the need for wild game was no longer significant. Except for those relatively few peoples who continued to hunt as a way of life, wild game ceased to be an important element in the human diet. In the New World, for a time game was a marketable commodity, and professional gunners profitably plied their trade until they ran out of wildlife to kill in the latter part of the nineteenth century. In our own time, however, it is difficult to understand why the pursuit of such wild game as remains continues to be carried on by North Americans of European descent, when the necessity no longer exists.

On a lovely African evening, Roger Peterson and I were driving our Land-Rover up the slopes of Mount Kenya on the way to our campsite. We chanced to pass the tents of a large, highly organized gunning safari. Spread round the campfire were the inert trophies — a pair of elephant tusks, a rhino head, the stretched hides of antelopes. That sight, with all its implications, reeks in my memory. In the twentieth century, African mammals such as elephant, sable antelope, bongo, greater kudu, and many more, have been reduced by nine-tenths. The romance of Darkest Africa is deeply involved with killing, as is the romance of those North American frontiers that yet remain.

The Judaeo-Christian progress tradition has come to be intimately involved with the profit motive, which in turn is intimately involved with man's impact on nature. The economics we know are Western phenomena; it was Max Weber, the turn-of-the-century sociologist, who observed that Western civilization is unique in that it supposes that its cultural trappings have "*universal* significance and value." Today, Western cultural traditions are dictating not only the course and destinies of those nations in which they originated, but also those of the "underdeveloped" or "developing" nations, which are in the process of being encouraged — one might even say coerced — to aspire to those values. None of these is more important than the economic system which is based on perpetual growth in the production and consumption of goods. Most of today's "under-industrialized" nations have been persuaded that in order to achieve the highest level of success, they must adopt the traditional Western approach to trade and commerce, and all the radical assumptions which underlie that approach. This has global environmental repercussions, as a potentially suicidal drain on non-renewable resources.

Resources were considered to be infinite and inexhaustible when Adam Smith pronounced in his 1776 *The Wealth of Nations* that, assuming that the object is to increase the national wealth, "this object will be most effectually secured by perfect industrial liberty." The consequences of perfect industrial liberty of access to resources have become conspicuous in our time. Smith's pronouncement became dogmatized, and a new body of faith – a creed – had emerged. Leave industry to its own devices, and all will be well. Such truths are universal, said Smith.

The universal truth of *laissez-faire* found an eager and willing marriage partner in Calvinism, and although a long series of people had conspired to bring the two together, the most generally recognized final matchmaker was Benjamin Franklin. Franklin had inherited all the Puritan virtues, and although the strictly religious context had largely evaporated by his time, all the fundamentals remained. As Weber says, "Honesty is useful, because it assures credit; so are punctuality, industry, frugality – and that is the reason they are virtues." These truths, among others, were held to be self-evident.

Weber's thesis, subsequently elaborated in R.H. Tawney's *Religion and the Rise of Capitalism*, was that there is an intimate and inseparable relationship between the asceticism of traditional Calvinism and the emergence of modern capitalist institutions. He points out that when asceticism "began to dominate worldly morality, it did its part in building the tremendous cosmos of the modern economic order. This order is now bound to the technical and economic conditions of machine production which today determine the lives of all the individuals who are born into this mechanism, not only those directly concerned with economic acquisition, with irresistible force." Machine production, of course, is not unique to capitalism.

George Grant remarks that "the public virtues [Franklin] advocates are unthinkable outside a Protestant ethos. . . . In 1968 Billy Graham at the Republican convention could in full confidence use Franklin in his thanksgiving for what the

Christian God had done for North America." The tradition of Dr. John Donne is long. We all remember Franklin's "God helps them that help themselves," his "Early to bed and early to rise . . ." and his "Plough deep while sluggards sleep." Ralph Waldo Emerson was able to say it somewhat more elegantly: "A creative economy is the fuel of magnificence." The propriety – indeed, the sanctity – of economic progress was not seriously questioned until our own time. With one notable exception.

John Stuart Mill, who died one hundred years ago, was wildly out of step with his time in one sense at least. Unable to accept that the wealth of nations was governed by immutable universal law, he chose to suggest the heretical notion that sheer material progress might not be the best of all possible goals.

With rare sensitivity and the intellectual courage of a Malthus, Mill suggested that when we cease our preoccupation with "the art of getting on" materially, there might emerge another kind of civilized life style. He questioned the desirability of population growth and economic growth. "I sincerely hope, for the sake of posterity, that they will be content to be stationary before necessity compels them to it. A stationary condition of capital and population implies no stationary state of human improvement. There would be as much scope as ever for all kinds of mental culture and moral social progress, as much room for improving the art of living. . . ." Mill, like Malthus, could not anticipate the technological revolution of the twentieth century, nor the enormity of the population crisis, but his insights were sound.

Stranger still were Mill's grave doubts about the desirability of "every rood of land brought into cultivation which is capable of growing food for human beings, every flowery waste or natural pasture plowed up, all quadrupeds or birds which are not for man's use exterminated as his rivals for food, every hedgerow or superfluous tree rooted out, and scarcely a place left where a wild shrub or flower could grow without being eradicated as a weed in the name of improved agriculture." It was Mill's persuasion that such events need not be inevitable. Things are subject to direction. Decisions could be made, and changes in trends could be implemented. He suggested that population growth could be curtailed. He was strikingly contemporary in his view that once a "stationary" state of society was achieved, there would be the means, through improved methods of distribution of products rather than the sheer quantity of production, for a reasonable degree of comfort for all.

Like Mill, Max Weber brought new dimensions to the examination of economic systems, dimensions which are of critical importance to today's environmentalist. Both were strongly influenced by social factors, and Weber was at pains to emphasize the plurality of causes in the evolution of economic tradition. Like the contemporary ecologist, he recognized the effects of a multiplicity of variable forces playing upon streams of happenings. No event was likely to have one simple origin. In a socio-economic system, as in an ecosystem, many influences are so intimately interrelated that to look for single motive causes is an empty exercise.

Weber did, however, identify a climate of acquisitiveness in Western European culture in which most men were inescapably trapped. He identified too an objective and indifferent (Judaeo-Christian) hierarchy of organization – the bureaucracy – which is a "mechanism founded on discipline," and has little to do with the free and open choice of the individual to determine his own destiny. Locked in the bureaucracy, a man goes on producing as the machinery dictates. This was considered by some, such as Thorstein Veblen, as the normal pattern. He believed that there is in man an innate desire to produce, and that the motive was simply an "instinct of work-manship," not profit. This tendency obviously played into the hands of the industrial system. That there is little of the innate or instinctual in human behaviour, Veblen could not know, but he was as modern as today in his observation of the "inordinate productivity of the modern machine process."

Inordinate productivity is what present environmental issues are concerned with — not capitalism or socialism. Environmental issues must not be confused with ideological issues. From the conservationist's point of view, there is no shred of difference between capitalist and socialist societies so long as both stand for inordinate industrial growth and productivity. Industrialization at an increasing rate is the goal of all of the super-powers and their satellites today, and industrialization (including growth in both production and consumption) is the grail of all forms of government in the "developed" world.

An article by Marshall I. Goldman, *The Convergence of Environmental Disruption*, makes the point: "By now it is a familiar story — rivers that blaze with fire, smog that suffo-cates cities, streams that vomit dead fish, oil slicks that blacken seacoasts, prized beaches that vanish in the waves, and lakes that evaporate and die a slow smelly death. What makes it unfamiliar is that this is a description not only of the United States but also of the Soviet Union." Goldman goes on to say that Lake Erie and Los Angeles have no corner on envi-ronmental degradation; the "debates and dilemmas are the same" for Lake Baikal and Tbilisi.

After a review of current environmental difficulties in Russia, Goldman concludes that "if the study of environmen-tal disruption in the Soviet Union demonstrates anything, it shows that not private enterprise but industrialization is the primary cause of environmental disruption. This suggests that state ownership of all the productive resources is not a cure-all. The replacement of private greed by public greed is not much of an improvement." In other words, there is little if any basis for the belief that differing economic systems have different degrees of impact on the environment so long as those systems are based on increasing industrialization. It is common nowadays to hear demands from various political persuasions and ideologists couched in ecological terms and directed toward the nationalization of industry. But there is a common bond between them all. All stand for the growth ethic and Calvinistic "progress." Those who are most vocal about the nationalization of Canadian non-renewable resources, for example, are operating from a bias which is equally at odds with sane environmental management as the system they are attacking. When nationalization of primary resources comes, as it inevitably must, it will be on grounds which are ecologically oriented. The grounds will not be ideological.

John Kenneth Galbraith reminds us that in the main (and, one would assume, despite their politics), economists are the natural allies of the industrial system. (*The New Industrial State* should be required reading for all environmental students.) Galbraith, who carries on the good iconoclastic tradition of J.S. Mill, and who is an equal embarrassment to his colleagues, has convincingly described the identification of the industrial system with the goals of society. If society chooses to limit itself to merely economic goals, says Galbraith, then it is only natural that the industrial system should dominate the state and that the state should in turn serve the ends of the industrial system. That is the bind into which contemporary Western society has got itself, with the industrial system having a virtual "monopoly of social purpose."

That the American style of industrialization today dominates the Western world and shapes the aims and ambitions of much of the non-Western world is one of those accidents of history, the product of a plurality of coincidences. What has happened is not at all unlike the result of any random shift of genetic material, influenced by chance environmental changes.

The plurality of causes, in their combination, became a phenomenon more combustible than any the world has ever seen. As Lewis W. Moncrief has described it, "America is the archetype of what happens when democracy, technology, urbanization, capitalistic mission, and antagonism (or apathy) toward natural environment are blended together." And running through the entire story, like a tight-drawn thread of the finest steel, is the spirit of John Calvin.

FIFTEEN

The POWER STRUCTURE

As a world species, we now find ourselves at the pinnacle of our ambitions. Man stands at the apex of an ancient and immensely complicated power structure over nature. The structure is man's, not God's; the ambition and the struggle – and the self-appointed mission to achieve it – were man's.

It has been a long and complex journey. The direction was initially the random result of evolutionary whim and chance environmental circumstance. Later, it caught the heady tide of tradition. Throughout, the pilgrimage has been accompanied by grave hardship, not only for men but also for other species and for a multitude of natural environmental types which perforce had to give way before the ultimate mammal. The journey has been an exemplary illustration of the opportunism which is characteristic of evolutionary events. Among the building blocks of the completed power structure there are the fossils of all the species that surrendered their living substance before us, and there are the shells of all the habitat types, large and small, that were eliminated or changed beyond recognition in the process. All this took place in a fraction of an instant of evolutionary time.

The journey of man has not been "purposeful." There were no built-in goals, any more than there are built-in directions for the evolution of a new kind of green heron or song sparrow. Random genetic and environmental events brought us to a stage at which our minds, and thus our cultures, took over. That, in point of time, was yesterday. The day *before* yesterday, changes were as always accidental and at random. The blinding speed at which cultural evolution proceeds tempts us to forget the slow process which is biological evolution. Cultural evolution represents the last twist on a curling metal "slinky-toy" of infinite length. But since we cannot see – or remember – beyond that final cultural spiral in the Series, we assume it was ever thus, and anything that happened in pre-historic times has no relevance to our behaviour.

Piece by piece, bit by bit, long before we could even formulate the notion, the power structure over nature grew and strengthened. The final "clincher" came with modern technology, which made possible the subjugation of America in two hundred years. The power and the mastery of Bacon and Descartes had become progress in the economic sense, and all had been sanctified by the Protestant church, most especially in its New World, Calvinist form.

As our minds are limited in the number of dimensions we can contemplate, it is useful to picture the power structure as a great latticework in a shape like that of the Eiffel Tower. Man stands at the top. Below him are all the nonhuman components of the biosphere, in descending order according to their relative value to man, which is the sole measure of their cosmic significance. Thus, man stands on Western technology, and Western technology stands on energy resources. Below energy resources there are non-renewable resources. The next stratum downward consists of those resources such as soils, forests, fish and wildlife which we used to call "renewable." We are now a very long way down on the hierarchical scale. Of less importance than renewable resources, and thus of less validity, are "non-economic" animal species. Below the animals which we consider valueless are those plants which have no economic "purpose." Further down is terrestrial wilderness. Of least importance is marine wilderness.

The very word "resource" trumpets Old Testament self-centredness — a thing nonhuman but useful to man. If it is not useful to man, it is not a resource. It is not, in fact, anything. Resources, on the other hand, have been specially provided as our manageable "harvest."

When the "harvest" principle first evidenced itself in such practices as wildlife management, it was a positive and beneficial concept. It emphasized that one should not, in taking wildlife, eat into capital — or at least not enough to endanger next year's production. The harvest principle is still used, however, in a time when we have learned that there is no harvestable surplus of anything, and there never was. A harvest is a planted, domesticated crop — a stand of wheat or field of tomatoes. For one species of living thing to regard another as a "crop" seems to be the very height of egocentricity and the depth of insensitivity. Before man was in North America to "harvest" the ducks and the balsam fir, the natural system did an exemplary job of feeding these things back into its cycles. If — and it is a big "if" — we can conceive of "harvestable surpluses" in nature, these went to sustain the other component species in the natural community. "Excess" ducks fed pike and prairie falcons, and excess balsam fir fed moose in the winter. Since there was no harvester, there was no harvest, and the wildlife community functioned sublimely well in their absence.

Harvest, resources, management — these concepts have no existence outside our self-centred minds. None is real, except when they are used as justification for further prosecution of the holy war on nature. Like the power structure itself, they are sustained by assumptions which are the result of our most ancient traditions.

Since it is clear that second only to man himself on the Great Scale of Perfection is his Western technology, we are obliged to look at a phenomenon which has been carefully explored by a long list of students in the present decade. It seems that a generalization one can make about technology today is its mindlessness. In the present day, technology proceeds for its *own* sake, not even for the human sake, in a succession of self-fulfilling prophecies. Technology makes something possible, then technology *must* go ahead and do it, regardless of necessity or consequence. As Buckminster Fuller says, "The horrifying truth is that, so far as much technology is concerned, no one is in charge."

Technology in its present form is something quite new. What was originally a kit of increasingly useful tools for helping us get things done has begun to proliferate for the sake of proliferation. Fuller reminds us that in both its capitalist and communist form, industrialization was directed to the advancement of material welfare: "Thus, for the technocrat, in Detroit as well as Kiev, economic advance is the primary aim; technology the primary tool. The fact that in one case the advance redounds to private advantage and in the other, theoretically, to the public good, does not alter the core assumptions common to both. Technocratic planning is *econocentric*." Technology is the inevitable and indispensable handmaiden of economic growth through industrial production. In today's world — if it is true that we are entering a post-industrial phase — technology having achieved its own momentum, proceeds in its own directions, irrespective of human requirements.

Many conservationists have come down hard on the "technocrats." It may be, however, that we blast the engineers and their accomplices somewhat too freely. One is not quite so Christlike as to intercede for their forgiveness, but it is obvious that they know not what they do. Who has been at pains to educate or even to encourage them otherwise? Who has troubled to inject even the slightest germ of comprehension of man *as* nature into schools of forestry, engineering, or architecture? Who has attempted to suggest to such institutions that there are ecological ground rules, much less ecological ethics? If technical men have so firmly welded and riveted together the supranatural tower of human dominance, then like all technical men they did so at someone else's behest. The fault has never been in our stars. It has been in our traditions.

The majority of our schools are still immured in traditions which are no longer merely irrelevant, but destructive. As Paul Ehrlich has said, "Our educational system has seriously straightjacketed the economic thinking of most of our citizens to the point where they are mesmerized by the axiomatic 'good' of growth." We have encouraged – even directed – students to look to growth for the ultimate satisfaction and to look to technology for the achievement of growth. The ecological innocence and indifference of a high proportion of graduates from most traditional schools is appalling. Our educational system is no breeder of environmental humility.

The power structure has become critical to the entire biosphere in our time because of the exponential increase in our *ability* to achieve the conquest of our environment. The goal has always been there; only the methods have changed. Modern tools have allowed us to accomplish what we would certainly have accomplished in earlier times, had we had the techniques.

It is true that our contemporary picture of "progress" is relatively recent; in its present form it dates back no further than the seventeenth century. But the blueprint for the tower of Babel existed long before that. Man and nature have been separate in men's minds since long before there was any notion of industrial progress. They were conceptually separate in a time when all things were seen as static and immutable. There was a pyramid even then. Teilhard's tree is at least as old as Aristotle, and power over nature goes back to faintly magical stirrings in a deep and lightless Neanderthal cave.

Although Charles Reich's description of today's elite (in *The Greening of America*) is more of an indictment than a portrait, it is impossible to disagree with him. The elite, he says, have been cruelly deceived into believing that "richness, satisfactions, joy of life, are found in power, success, status, acceptance, popularity, achievements, rewards, excellence, and the rational, competent mind." They want "nothing to do with dread, awe, wonder, mystery, accidents, failure, helplessness, magic." Such people, who make up the contemporary "establishment," Reich says, have produced a society that is "the image of its own alienation and impoverishment." He goes on to remark that there is an "imperative logic" behind the destruction of nonhuman nature as practiced by the Corporate State, because the state "draws its vitality by a procedure that impoverishes the natural world." We have always had that imperative logic; it has been part of the evolving power structure since prehistory. Today, our technical efficiency has matched our aspirations. The danger is that the technological machine, in its mindless acceleration, may now run away from *all* of nature — which includes ourselves.

I have entertained a vision of a sea cucumber — a marine invertebrate belonging to the animal phylum called echinoderms, the spiny-skinned ones. Echinoderms are usually characterized by five-part symmetry, as with the legs of a starfish or the segments of a sea urchin. Like the others in its phylum, and unlike all other animals in the world, the sea cucumber's vascular system is filled not with blood, but with water. It is a sluggish and slow-moving animal which lies more or less immersed in muck at the bottom of the ocean. It takes sludgy water in through its mouth, extracts whatever nutrients the water contains, and voids water and unusable sand at the other end. The sea cucumber, as it blindly pulses on the ocean floor, does not select its food; it ingests whatever happens to be in front of it in the watery sediment, and it voids whatever it cannot use.

There is a striking analogy between the sea cucumber and the insensate ingesting, processing and voiding of modern industry. Like the echinoderm, industry does not care a hoot what it takes in at one end, and cares even less about what it deposits at the other, so long as it can extract something of value in the process. Squeeze a sea cucumber and you get water, not blood. Squeeze industry and you do not get blood either. Squeeze the echinoderm a little harder, and the animal will undergo a violent paroxysm in the course of which, according to Rachel Carson, it will "hurl the greater part of its internal organs through a rupture in the body wall. Sometimes this action is suicidal, but often the creature continues to live and grows a new set of organs." Squeeze industry hard enough and it, too, will behave as though you are killing it.

Since neither sea cucumber nor industry appears to take the slightest interest in what flows in the front door and flows out the back, so long as it is nourishing in the process, one can take a moment of frivolity to indulge in the alchemist's fantasy. What would the organism need to ingest in order to cause it to excrete pure gold? Or perhaps pure air. All that would be needed is the right formula. It must be said of the sea cucumber that it voids almost nothing but the finest and cleanest sand. Moving sand from place to place is an innocuous activity which may even be beneficial in the long run. The benefits of most industrial processing today, apart from those involved with basic necessities, is another question.

However, like the sea cucumber, the industrialist is not exercising willful choice. He does the only thing he knows how to do, and he does it because he has always done it. He differs from the echinoderm only in that he believes what he does to be right and proper. Both "do their thing" because that is what they have evolved to do within the context of their special environments. Both would have disappeared onto the slag-heap of misfits long since had their environments not permitted them to perpetuate themselves.

No man is so far removed from nature as the liberally educated humanist, because the cosmos centres on his mind, and the mind of man is the measure — and the envelope — of all things. This is the man to whom, as Paul Shepard says, nature is either natural resources or scenery — or else it is that endlessly fascinating garden of delights called "human nature." The run-of-the-mill humanist is incredibly ignorant of, and thus indifferent to, his biological context; and he is even somewhat reluctant to be reminded of it. The liberal humanist is dangerous to the biosphere, and thus to mankind, because by and large he is the leader, the educator, the opinion-maker. Even in a money-oriented society, his influence exceeds that of the technocrat or the industrialist. His "culture" grants him access to seats of power and influence. He is the key to the entire supranatural pyramid, because he is ancient anthropocentricity in its most highly developed form.

The humanist is the very picture of tolerance and compassion toward nature. He supports "humane" societies, he takes his children to the zoo to see the amusing antics of captive "lower" animals, and to the museum to see curiosities from the farthest corners of man's world. He is interested in everything, because he has a cultivated mind. He is gentle, genteel, and humanitarian. Unlike men in trade, commerce and the technical professions, the liberal humanist has been at an evolutionary standstill for over two thousand years.

That is a regrettably long time, in view of the potential speed of cultural evolution by comparison with the more deliberate pace of biological processes. While most other aspects of our culture were undergoing remarkable transformations and developments, the traditional humanities remained locked in antiquity – whether in a Romantic fixation on things past or in a protective shell is open to opinion.

Although one is chagrined at the continuing failure of the humanities to enter the contemporary arena of environmental issues, and at their apparent insistence on maintaining the rigid integrity of the conceptual separation of man and nature, it is nonetheless in the humanities that we must concern ourselves.

If "human ecology" is ever to emerge as a definable body of knowledge or area of investigation, it is far more likely to emerge from the humanities than from any of the hard or soft sciences. Man is the cultural animal. Culture created the power structure over nature, and only in culture is the blueprint for its dismantlement.

The GRACEFUL CONTINUITY

Except for rare aberrations, the humanistic tradition has never been reconcilable with the realities of living on Earth. There is a critical need for a dispassionate and thoughtful review of the basic assumptions that underlie our literature, philosophy, and arts. Only in this way can we begin to pry open the shackles of closed thinking that have plagued Western culture for so many centuries. This is where artificial selection comes in. We develop fancy pigeons by selecting for breeding those birds with the attributes we desire. The others we suppress. The same can be done with component parts of our culture. But before we can select those components for propagation or rejection, we must be able to identify and evaluate them within the complex "ecosystem" which is our collective accumulated thought.

Values have been variously defined, but a common element appears to be the perception of some degree of merit, or at least of usefulness, in the eye of the possessor of the value in question. Values are also defined as qualities, standards, or even principles, which are considered by the possessor as desirable, or at least worthwhile. These perceptions and evaluations are of course no more than indications of what goes on inside the head of the individual beholder, or valuer. They have no intrinsic "value," no intrinsic reality. The reality is in what we *do* on the basis of our values. In a sense, a value is at the same time a *motivation* to achieve the satisfaction and pleasure one "values." Thus a value becomes a *reason*. Like energy and matter, satisfaction and motive become at once the same and inseparable.

Institutions are, among other things, reflections or manifestations of sets of currently prevailing values. An obvious example is the GNP. The GNP is an institution; even today, relatively few people question its value as a measure of progress. Progress itself is an institution. The conceptual separation of man and nature, and the conceptual power structure over nature, are institutions which reflect prevailing values. The power structure itself is built on other institutions — religious, philosophical, economic, political, humanistic. These are traditional "givens"; they are reflections of subjective value systems making up the largely artificial "construct" which is contemporary Western culture.

Institutions such as the self-perpetuating profit system and the gospel of individual freedom are additional evidence of the value systems we have inherited. They are firmly ingrained in us; to change them would be difficult. The acceptance of any new set or sets of values would involve a change in "mental set" of a magnitude and profundity which would only be comparable to the intellectual and emotional "conversion" implicit in certain kinds of religious experience. Certainly it would have to be on a grand scale. But it could happen: there are precedents.

Even very radical shifts in value systems are possible if there are good reasons for them. The fact that values are often their own reasons can help to make the transition smoother.

One thinks of times of national emergency, such as war, when the value of individual freedom becomes relatively unimportant in the face of the greater social purpose. Values such as the sanctity of human life evaporate on the instant. Even the value of profit-making (to at least some extent) and that of fair competition are secondary to the war production effort. The reason is the prosecution of the war, and the prosecution of the war becomes the new value. The institution of democracy is temporarily, and quite voluntarily, suspended.

There can also be shifts in value systems when there are reasons such as the relative abundance or scarcity of commodities or even objects of interest. Society is quite prepared to value the whooping cranes or the blue whales when it is informed that there are so very few of them. It is civilized to value things that are in short supply. If they are few, they must be valuable. Society is somewhat less inclined to value things such as raccoons or redwinged blackbirds because there are still lots of them.

There have been changes in institutionalized values and value systems when developments in popular thought have declared them to be irrelevant or archaic. In our own time we are seeing at least some changes in the values formerly attached to such institutions as wedlock, motherhood, militarism, conspicuous consumption, the old-time religion, haircuts, bras and neckties. If people feel better without these accoutrements, then their absence is the new value. In their place have been substituted such things as chronic skepticism, blue jeans, compulsive sharing, and whatnot. Whether the values of "Consciousness III" will persist is of course unknown. What *is* known is that spiritual bankruptcy is in institutions, not in people.

In their natural environments, living beings face an infinity of survival problems – food shortages, predators, diseases, competitors, population stresses, and so on. The gravest problem they now face, however — man's self-appointed supremacy over them — is itself strangely like an ecosystem. It has a vast and complicated array of interlocking components. All components interact with each other. All pull and push and pulse with rhythmic tensions, as in a dynamic web of nature. The conceptual power structure is so old, its foundations so embedded in time and tradition, that it has become a very real community of mutual self-interest, sustained and reinforced by its interacting components. Its homeostasis is maintained by the constant energy exchanges and mutual feedbacks between myth and superstition, religion and philosophy, science and technology, politics and economics, fear and arrogance. All, for their symbiotic survival, have depended upon all the others in the creation of a complicated web of interdependent strands which has now become a conceptual monolith of terrifying scope — and stability.

But the very fact of the structure's stability means it is not static. Like any other ecosystem, it is in constant flux. Also, as in the web of any other ecosystem, pressure on one strand is transmitted in greater or lesser measure to all the others. Pull on the strand of traditional, classical economics, and you get immediate political oscillations. Gently nudge the strand of education, and you get sympathetic responses from unexpectedly distant branches of religion. Pluck the thread of progress, and violent are the paroxysms throughout.

As any naturalist knows, the quickest and neatest way to destroy an ecosystem is to simplify it, to reduce its complexity and thus to short-circuit the equilibrium maintained by the mutual interdependence of its component parts. Perhaps the traditional, cultural, institutional, conceptual eco-construct can be decomplexified by our deliberate manipulation — by the exercise of our conscious choice.

Intervention in its workings will require degrees of courage, sacrifice, imagination and generosity which have not frequently been displayed in the course of man's relationship with his environment. One hesitates to predict whether we will be willing to undertake it. The destruction of the power hierarchy over nature will require a shift in attitudes more profound than we can presently imagine.

The process of simplification or decomplexification will be drastic. Suppose one were to elect to have an initial go at the "rights of man" — the God-given rights of man the individual and man the species. Suppose it were feasible to actually remove some of those rights, one after the other. The consequences might be astonishing. Let us think of such traditions as every man's right to a plot of land with a house on it, the land to do with as he will. A man can do as he wishes with his own property — his trees, his water, his soil, and so on. We now know, however, that a man can no longer pollute as he pleases. He is controlled in that. He will soon be controlled much more stringently in his use of the soil itself. A man should no more be allowed to own the living soil than he now owns the air he breathes. Environmental forces are already eroding traditional "rights."

Then there is the right to have children. Suppose people were no longer permitted to reproduce beyond the replacement level. Replacement means one adult, one child — zero population growth. In more practical terms, it means two children per female human being. Is the imagined right to have as many children as one pleases really in the best interest of all mankind? One might go further, and say that there are some people who do not have the right to have children at all. There are other people who are quite capable of producing intelligent, strong, beautiful children, but whose right to *raise* them might be gravely questioned. I know children of immense promise who are being destroyed by parents of eighteenth-century persuasions. The environmentalist must look hard at traditional human freedoms.

There are other "rights" such as the imagined right of men to kill nonhuman animals for amusement. Clearly the environment itself will deal with this tradition, simply as the effect of men having joyfully massacred so many ducks, geese, rhinos, elephants and Cape buffalos that there will not be enough of them to go round. A similar end will come to the fashion industry's apparent determination to exploit to the bitter end the final stocks of leopard, tiger, jaguar, and the rest.

What of the more fundamental, unquestioned rights of man the species? The right to populate at will must certainly be removed, either by our conscious choice or by a natural backlash on the part of the biosphere itself. The right to dominate animals of other species, and to dominate landscapes, will not be subverted as readily. Other beings, as species and as landscapes, do not have the "clout" of the combined forces of the biosphere. But that right, too, will disappear. It will be a sad process, for we will not give up the right to dominate without a struggle – a struggle which will cost both human and nonhuman nature exorbitantly.

It will not be in our best interests to allow the environment to dismantle our conceptual power structure for us. In such an eventuality, cosmic forces would make life devastating. We should not expect the environmental counterattack to be nearly so dramatic or spectacular as the ancient vision of the Apocalypse, but it would be equally disastrous. Because it would not be sudden, it would be even more agonizing.

It has never been difficult for us to see the necessity for and to devote abundant energy to the inflicting of our ideologies on people who because of their circumstances are not privy to our insights. It should not be difficult for those of us who have been privileged to receive "the word" about the deteriorating *human* condition to inflict that word, by one means or another, on those who have not got it. We have the evidence. Unlike redemption, or any political "ism," the environmental crusade has plenty of hard fact to back it up. There is no mystery about it. The threat of hellfire is no longer veiled; there are millions of starving people and there are billions of mutilated nonhuman beings to bear witness to the dislocation of the biosphere.

One of the saddest ironies is human adaptability. As a mammal, man has an internal heat-regulating system and can live in a variety of climates. As a primate, man is a sort of generalist and can live in a wide variety of habitat types. As a technologist, man can live virtually anywhere. More than that, however, is man's psychic adaptability. We can persuade ourselves of almost anything, however fanciful – such as our separation from nature. Worse, we can get used to living under the most trying conditions. People suffered abominably in Hitler's concentration camps, but a good many made a sufficient psychic adjustment. People live today in indescribable filth and squalor, without enough food, without medical assistance, without hope. But their psychic adaptability allows them to *live*, and millions of them live long enough to beget millions more. People in large industrial cities get used to levels of air pollution which choke and nauseate new arrivals from the "wide open spaces." We get used to these things, just as we get used to the discomforts of aging, because they are gradual developments, and we are rarely aware of what is happening until it has happened. Adaptability can be insidious.

Adaptability can produce startling results for a species – and in the psychic sense, even for an individual – but one thing that adaptability can never do is guarantee immortality. Immortality has yet to be demonstrated in the cosmos, and it has yet to be demonstrated on Earth. Certainly it will not be demonstrated in the lifetime of the species man. No cell, nor individual, nor species, is immortal. There is comfort in that fact. Eternal life, like life imprisonment, would be the cruellest punishment that could be inflicted on a living being.

One wonders why the dream of immortality has always been so important to us as individuals, as societies, as cultures, and as a species. The naturalist is not all that "hung up" on immortality. I made my peace with that issue years ago (and it *is* a peaceful thought). Saul Alinsky was recently quoted in *Playboy*: "Once you accept your own mortality on the deepest level, your life can take on a whole new meaning."

Earth will continue in its orbit for the balance of its own lifetime regardless of present and future human activities. Life processes and their manifestations may be simplified by contemporary standards, with organisms of a somewhat lesser degree of overt complexity of organization filling the initial void which our last convulsions will have left. But the overall background complexity – the programming – will remain. Some of Earth's matter may by that time have been released as energy by nuclear devices, but only a minuscule amount. Most of the matter that is here today will still be here. The essential atoms may be in somewhat different molecular combinations, but since nothing goes away, all will be here. In whatever form we leave them, the basic elements will eventually sort themselves out into new, workable structures which will be adapted to the new, posthuman biosphere.

Man *might* take with him more than he should of the necessities for posthuman life on Earth. The ultimate blasphemy would be man's total destruction of the materials and possibilities of the life process. This, I reassure myself, is impossible. Man will be prevented from committing this final and ugliest of acts by the sheer resilience of the biosphere. It is beyond our capacity to destroy everything. In fact, it is beyond our capacity to destroy *anything* – merely to rearrange its parts.

The appeal of mortality is in the graceful continuity which it makes possible. Can you picture anything more dull and depressing than an inert, static, unchanging world, with all things just as they are today? The glory and the grandeur are in the flux, the constant shifting which is made possible by the combination, decomposition and recombination of the basic building materials of life. Though I do not expect that I shall be reborn directly as a crocus, I know that one day my atoms will inhabit a bacterium here, a diatom there, a nematode or a flagellate – even a crayfish or a sea cucumber. I will be here, in myriad forms, for as long as there are forms of life on Earth. I have always been here, and with a certain effort of will, *I can sometimes remember.*

We attempt to distinguish between past, present, and future when no such distinction is possible. Those spinning particles that at this instant are John Livingston have always been around, somewhere, and always will be. At this instant, quite by chance, they happen to be in their present configuration. We attempt to distinguish between plants and animals, but the harder we press our determination to differentiate between them, the fuzzier the distinction becomes. We attempt to distinguish between living and nonliving, and then we stumble on the virus. We attempt to distinguish between man and other living beings, when there is no distinction. Differentiation is the cruellest block to comprehension of the oneness of being.

Oneness and the notion of timelessness are hard on mental images of purposes and goals, on pictures of causes and effects, on visions of destinies. The ancient Greeks knew better about these things than did Francis Bacon, and they knew much better than the contemporary decision-maker raised in the Western humanistic tradition. Not only did the Greeks know about oneness, they knew also about *hubris*. Alexander Pope knew about the error of reasoning pride, and Shakespeare about the relative significance of the poor posturing player.

Time is fast running out for the dismantling of the institutions which have kept us so grimly locked in step with "progress." There is even less time for reflection on the merits of the traditional components of our culture which have brought us — and all of nature — to the present point of departure. A point of departure it is, either from the narrow and egocentric cultural course we have adopted, or premature departure from the blue planet itself. If we are not yet capable of identifying the specific threads in the fabric of our beliefs which have sustained the entire tapestry upon which the myth of human dominance is emblazoned, then it may be too late already.

The hope for survival of nonhuman nature is dim. There is a familiar scenario. As conditions worsen for human populations — as they will, initially, in underprivileged parts of the world — every ounce and erg of our most refined technological skills and energies will be brought into play to extract from Earth and its nonhuman inhabitants the basic ingredients for human survival. We will first destroy all of the larger animals, either for meat or because they compete with us for space, together with those which may be intolerant of our activities because of their specific natural specializations. Extinction of nonhuman species, without replacement, will continue at an accelerating rate, until the only nonhuman living beings remaining will be those who are willing to share their squalor with us — rats, gutter curs, and parasites and micro-organisms which thrive in times of environmental dislocation.

Our capacity for seeing into the future is limited — and we do not want to know about futures of that kind. We withdraw behind the opaqueness of closed imaginations and familiar fancies. We acknowledge that, yes, the situation is bad, but human ingenuity, creativity, enterprise and goodwill can overcome all difficulties.

While we should be unravelling the threads of tradition, we are weaving ever more elaborate curtains of rationalization. Every avenue of questioning closed off is another route to intellectual and spiritual freedom barricaded forever.

There is no engineering answer to a problem created by culture. The worst in humanistic ways of thinking opened and kept open the conceptual man/nature dichotomy, and only the mature wisdom and insight that characterize the best in the natural philosophic tradition can mend it.

BIBLIOGRAPHY

Adamson, George, *Bwana Game*. London: Collins & Harvill Press, 1968.

Ardrey, Robert, *African Genesis*. New York: Atheneum, 1961.

—— *The Territorial Imperative*. New York: Atheneum, 1966.

—— *The Social Contract*. New York: Atheneum, 1970.

Babbitt, irving, *Rousseau and Romanticism*. Cleveland, Ohio: Meridian Books, 1955.

Bates, Marston, *The Forest and the Sea*. New York: Random House, 1960.

Bernal, J.D., *Science in History* (4 vols.). Harmondsworth, Middlesex: Pelican Books, 1969.

Berrill, N.J., *Man's Emerging Mind*. New York: Dodd, Mead & Company, 1955.

—— *You and the Universe*. New York: Dodd, Mead & Company, 1958.

Bertram, Colin, "Man Pressure." *Oryx*, 7 (2-3): 97-101, August, 1963.

Biggar, H.P. (ed.), *The Works of Samuel de Champlain*. Toronto: The Champlain Society, 1936.

Brower, David (ed.), *Wilderness: America's Living Heritage*. San Francisco: Sierra Club, 1961.

Buchan, John (ed.), *A History of English Literature*. London: Thomas Nelson and Sons, Ltd., 1923.

Buettner-Janusch, John (ed.), *Evolutionary and Genetic Biology of Primates* (2 vols.). New York: Academic Press, 1963.

Burrill, Donald R. (ed.), *The Cosmological Arguments*. Garden City, New York: Doubleday (Anchor Books), 1967.

Carson, Rachel, *The Edge of the Sea*. Boston: Houghton Mifflin Co., 1955.

Chard, Chester S., *Man in Prehistory*. New York: McGraw-Hill, 1969.

Clark, Kenneth, *Civilisation*. London: British Broadcasting Corporation and John Murray, 1969.

Clement, Roland C., "The Ethics of Survival." *The Ohio Journal of Science*, 69 (1): 15, January, 1969.

Coale, Ansley J., "Man and His Environment." *Science*, 170: 132-136, 1970.

Commoner, Barry, *The Ecological Facts of Life*. U. S. National Commission for UNESCO, 1969.

—— *Science and Survival*. New York: Viking Press, 1967.

Dansereau, Pierre, *The Hope of Human Ecology*. Canadian Commission for UNESCO, July, 1969.

Darling, Frank Fraser, *Wilderness and Plenty*. New York: Ballantine Books, 1970.

Deevey, Edward S., Jr., "The Human Population." *Scientific American*, 203 (3): 195-204, September, 1960.

Disch, Robert (ed.), *The Ecological Conscience*. Englewood Cliffs, New Jersey: Prentice-Hall, Inc., 1970.

Dorst, Jean, *Before Nature Dies*. Boston: Houghton Mifflin Co., 1970.

Dubos, René, *The Torch of Life*. New York: Simon and Schuster, Inc., 1962.

—— *So Human an Animal*. New York: Charles Scribner's Sons, 1968.

—— *Reason Awake*. New York: Columbia University Press, 1970.

Ecology Action East, "The Power to Destroy, the Power to Create." *Rat*, January, 1970.

Edberg, Rolf, *On the Shred of a Cloud*. University, Alabama: University of Alabama Press, 1969.

Ehrlich, Paul R. and Anne H., *Population Resources Environment*. San Francisco: W. H. Freeman and Company, 1970.

Eiseley, Loren, *The Immense Journey*. New York: Random House, 1957.

Elder, Frederick, *Crisis in Eden*. Nashville, Tennessee: Abingdon Press, 1970.

Elliott, J.H., "Spanish Holocaust." *The New York Review of Books*, November 5, 1970.

Falk, Richard A., *This Endangered Planet*. New York: Random House, 1971.

Firsoff, V.A., *Life, Mind and Galaxies*. London: Oliver and Boyd, Ltd., 1967.

Forstner, Lorne J., and Todd, John H. (eds.), *The Everlasting Universe*. Lexington, Mass.: D. C. Heath and Company, 1971.

Frazer, J.G., *The Golden Bough*. London: Macmillan, 1922.

Fuller, R. Buckminster, *Utopia or Oblivion: the prospects for humanity*. New York: Bantam Books, 1969.

—— *Technology and the Human Environment*. U. S. Senate Subcommittee statement, March 4, 1969.

Galbraith, John Kenneth, *The New Industrial State*. Boston: Houghton Mifflin Co., 1967.

Goldman, Marshall I., "The Convergence of Environmental Disruption." *Science*, 170: 37-42, 1970.

Goodman, Paul, "Can Technology be Humane?" *The New York Review of Books*, November 22, 1969.

Grant, George, *Technology and Empire*. Toronto: House of Anansi, 1969.

Hardin, Garrett, *Nature and Man's Fate*. New York: Holt, Rinehart and Winston, Inc., 1959.

Heilbroner, Robert L., "Ecological Armageddon." *The New York Review of Books*, April 23, 1970.

Hewes, Gordon W., "Food Transport and the Origin of Hominid Bipedalism." *American Anthropologist*, 63: 687-710, 1961.

Hockett, Charles F., and Ascher, Robert, "The Human Revolution." *Current Anthropology*, 5(3): 135-168, 1964.

Hopkins, David M. (ed.), *The Bering Land Bridge*. Stanford, California: Stanford University Press, 1967.

Howell, F. Clark, and the Editors of Time-Life Books, *Early Man*. New York: Time, Inc., 1965 (revised 1968).

Howells, William W., "Homo Erectus." *Scientific American*, 215: 46-53, November, 1966.

Hoyle, Fred, *The Nature of the Universe*. Harmondsworth, Middlesex: Pelican Books, 1963.

Huxley, Julian, *Essays of a Biologist*. London: Chatto and Windus, 1923.

Jackson, Wes, *Man and the Environment*. Dubuque, Iowa: William C. Brown Company, 1971.

Jastrow, Robert, *Red Giants and White Dwarfs*. New York: Harper and Row, 1967.

Josselyn, John, *New-Englands Rarities*. London: G. Widdowes, 1672.

Kruuk, Hans, "Clan-system and feeding habits of spotted hyenas (*Crocuta crocuta*)." *Nature*, 209 (5029): 1257-1258.

Kummer, Hans, *Primate Societies*. Chicago: Aldine-Atherton, 1971.

Laing, R.D., *The Politics of Experience*. Harmondsworth: Penguin Books, 1967.

Leakey, R.E.F., "Fauna and Artefacts from a New Plio-Pleistocene Locality near Lake Rudolf in Kenya." *Nature*, 226, April 18, 1970.

Leopold, Aldo, "The Conservation Ethic." *Journal of Forestry*, October, 1933.

—— *Sand County Almanac*. New York: Oxford University Press, 1949.

Livingston, John A., *Canada*. Toronto: Natural Science of Canada Ltd., 1970.

—— "Man and His World: A Dissent," in *Wilderness Canada*, Toronto: Clarke, Irwin, 1970.

—— "Neo-Nationalism and Ecological Independence – A Personal View." *Ontario Naturalist*, 8(4), December, 1970.

Lovejoy, Arthur O., *The Great Chain of Being: a study in the history of an idea*. Cambridge, Mass.: Harvard University Press, 1936.

—— *Essays in the History of Ideas*. Baltimore, Maryland: The Johns Hopkins Press, 1948.

MacNeill, J. W., *Environmental Management*. Ottawa: Information Canada, 1971.

Margolis, John, "Land of Ecology." *Esquire*, March, 1970.

Martin, P. S., and Wright, H. E., Jr. (eds.), *Pleistocene Extinctions: the search for a cause*. New Haven, Connecticut, and London: Yale University Press, 1967.

Marx, Leo, *The Machine in the Garden*. New York: Oxford University Press, 1964.

Matthiessen, Peter, *Wildlife in America*. New York: The Viking Press, 1959.

Mech, David L., *The Wolf*. Garden City, New York: The Natural History Press, 1970.

Merton, Thomas, "The Wild Places." *The Center Magazine*, July, 1968.

Moncrief, Lewis W., "The Cultural Basis for our Environmental Crisis." *Science*, 170: 508-512, 1970.

Montagu, M.F. Ashley (ed.), *Man and Aggression*. New York: Oxford University Press, 1968.

Morris, Desmond, *The Naked Ape*. London: Jonathan Cape, 1967.

——— *The Human Zoo*. London: Jonathan Cape, 1969.

Mumford, Lewis, "Closing Statement." *Future Environments of North America*, F. Fraser Darling and John P. Milton (eds.), The Conservation Foundation, 1966.

McHarg, Ian L., "Values, Process, and Form." *The Fitness of Man's Environment*, Washington, D. C.: The Smithsonian Institution Press, 1968.

McKinley, Daniel, "The New Mythology of 'Man in Nature.'" *Perspectives in Biology and Medicine*, 7(1): 93-105, Autumn, 1964.

Nash, Roderick, *Wilderness and the American Mind*. New Haven, Connecticut: Yale University Press, 1967.

Neill, Wilfred T., *The Last of the Ruling Reptiles*. New York: Columbia University Press, 1971.

Odum, E.P. and H.T., *Fundamentals of Ecology*. Philadelphia: W.B. Saunders Co., 1959.

Odum, Eugene P., *Ecology*. New York: Holt, Rinehart and Winston, Inc., 1963.

Odum, Howard T., *Environment, Power, and Society*. New York: John Wiley and Sons, Inc., 1971.

Ohlin, Goran, *Historical Outline of World Population Growth*. Background paper for United Nations World Population Conference, Belgrade, 1965.

Olsen, Jack, *Slaughter the Animals, Poison the Earth*. New York: Simon and Schuster, 1971.

Orenstein, Ronald I., "Tool-use by the New Caledonian Crow (*Corvus moneduloides*)." *Auk*, 89(3): 674-676, July, 1972.

Paddock, William and Paul, *Famine 1975!* Boston: Little, Brown and Company, 1967.

Potter, Frank M., Jr., "Everyone Wants to Save the Environment." *The Center Magazine*, March, 1970.

Prescott, W.H., *The Conquest of Mexico*. London: J. M. Dent & Sons, 1901.

—— *History of the Conquest of Peru*. London: J. M. Dent & Sons, 1908.

Prince Philip, Duke of Edinburgh, and Fisher, James, *Wildlife Crisis*. New York: Cowles Book Company, Inc., 1970.

Randall, John Herman, Jr., *Aristotle*. New York: Columbia University Press, 1960.

Reich, Charles A., *The Greening of America*. New York: Random House, 1970.

Reinow, Robert and Leona Train, *Moment in the Sun*. New York: Ballantine Books, 1967.

Romer, Alfred S., *Man and the Vertebrates* (2 vols.). Harmondsworth, Middlesex: Penguin Books, 1933 (revised 1954).

Schaeffer, Francis A., *Pollution and the Death of Man: the Christian view of ecology*. Wheaton, Illinois: Tyndale House Publishers, 1970.

Schaller, George B., *The Year of the Gorilla*. Chicago: The University of Chicago Press, 1964.

—— *The Deer and the Tiger*. Chicago: The University of Chicago Press, 1967.

Schaller, George B., and Lowther, Gordon R., "The Relevance of Carnivore Behavior to the Study of Early Hominids." *Southwestern Journal of Anthropology*, 25(4), Winter, 1969.

Schweitzer, Albert, *Out of My Life and Thought*. New York: Henry Holt & Co., Inc., 1949.

Sears, Paul B., "The Inexorable Problem of Space." *Science*, 129: 9-16, 1958.

Shepard, Paul, and McKinley, Daniel (eds.), *The Subversive Science*. Boston: Houghton Mifflin Co., 1969.

Slobodkin, Lawrence B., "Aspects of the Future of Ecology." *BioScience*, 13(1): 16-23.

Snyder, Gary, "Poetry and the Primitive." *Earth Household*, 1967.

Soule, George, *Ideas of the Great Economists*. Toronto: Mentor Books, 1952.

Spencer, Roscoe, "Individual and Species: Biological Survival." *The Humanist*, 3: 155-161, 1958.

Stalker, A. Mac S., "Geology and Age of the Early Man Site at Taber, Alberta." *American Antiquity*, 34.(4): 425-428, October, 1969.

Storer, John H., *Man in the Web of Life*. New York: Signet Books, 1968.

Storr, Anthony, *Human Aggression*. Harmondsworth, Middlesex: Penguin Books, 1968.

Teilhard de Chardin, Pierre, *The Phenomenon of Man*. New York: Harper and Row, 1959.

—— *Man's Place in Nature*. New York: Harper and Row, 1966.

Thomas, William L., Jr. (ed.), *Man's Role in Changing the Face of the Earth* (2 vols.). Chicago: University of Chicago Press, 1956.

Toffler, Alvin, *Future Shock*. New York: Random House, 1970.

Trevelyan, G.M., *English Social History*. London: Longmans, Green, 1942.

van Lawick-Goodall, Hugo and Jane, *Innocent Killers*. London: Collins, 1970.

van Lawick-Goodall, Jane, *In the Shadow of Man*. London: Collins, 1971.

Velikovsky, Immanuel, *Worlds in Collision*. New York: Doubleday, 1950.

Washburn, Sherwood L. (ed.), *Social Life of Early Man*. Chicago: Aldine Publishing Company, 1961.

Weber, Max, *The Protestant Ethic and the Spirit of Capitalism*. New York: Charles Scribner's Sons, 1958.

Weisberg, Barry, "The Politics of Ecology." *Liberation Magazine*, January, 1970.

White, Lynn, Jr., "The Historical Roots of Our Ecologic Crisis." *Science*, 155: 1203-1207, 1967.

Whiteside, Thomas, *Defoliation*. New York: Ballantine Books, 1970.

Willey, Basil, *The Eighteenth-Century Background*. Harmondsworth, Middlesex: Penguin Books, 1962.

Wright, Louis B. (ed.), *Newes from the New-World*. Companie of the Friends of the Huntington Library, 1946.

Wynne-Edwards, V.C., *Animal Dispersion in Relation to Social Behavior*. London: Oliver and Boyd, 1962.

—— "Self-Regulating Systems in Populations of Animals." *Science*, 147: 1543-1548, 1965.